Library of
Davidson College

THE BULGARIAN ECONOMY IN
THE TWENTIETH CENTURY

THE BULGARIAN ECONOMY
In the Twentieth Century

JOHN R. LAMPE

ST. MARTIN'S PRESS
New York

© 1986 John R Lampe
All rights reserved. For information, write:
St. Martin's Press, Inc., 175 Fifth Avenue, New York, NY 10010
Printed in Great Britain.

First published in the United States of America in 1986
ISBN 0-312-10785-4

Library of Congress Cataloging in Publication Data
Lampe, John R.
  The Bulgarian economy in the 20th century.

  Bibliography: p.
  Includes index.
  1. Bulgaria—Economic conditions. 2. Bulgaria—
Economic policy. I. Title. II. Title: Bulgarian economy in the twentieth century.
HC403.L36   1986   330.9497'703   85-14549
ISBN 0-312-10785.4

CONTENTS

List of Tables and Maps
Preface
Editor's Introduction   1
Introduction   13
  1. Initial Growth, 1878–1918   19
  2. Recovery in the 1920s   49
  3. Isolation in the 1930s   78
  4. The Second World War   105
  5. Communist Revolution, 1944–1947   121
  6. The First Five-Year Plans   139
  7. Industry and Agriculture since 1960   156
  8. Foreign Trade and Domestic Living Standards   177
  9. Economic Reforms since 1960   199
Conclusion   223
Select Bibliography   231
Index   237

LIST OF TABLES AND MAPS

1.1	Gross Social Product in 1911	25
1.2	Real Exports per capita, 1886–1910	25
1.3	Distribution of Cultivated Land, 1897 and 1911	26
1.4	Ownership of Private Rural Land, 1897 and 1908	28
1.5	Main Financial Indicators, 1886–1911	33
1.6	Distribution of Bank Assets in 1911	33
1.7	Industrial Output, Capital and Firm Size in 1911	36
2.1	Grain Production and Consumption, 1921–30	53
2.2	Composition of Exports, 1907–30	53
2.3	Direction of Foreign Trade, 1906–30	54
2.4	Sources of Population Change, 1906–40	55
2.5	Main Financial Indicators, 1911–30	61
2.6	Loan Value by Bank Type, 1911 and 1928	67
2.7	Structure and Growth of Industry, 1911–31	69
3.1	Indices of Crop Output, Area and Yield, 1926–38	85
3.2	Foreign Trade Balances, 1926–38	90
3.3	Direction of Foreign Trade, 1929–39	90
3.4	Growth of Net Industrial Output, 1921–37	96
3.5	Number and Size of Industrial Firms in 1937	96
4.1	Indices of Agricultural Production, 1939–44	111
4.2	Foreign Trade and Inflation Indices, 1939–44	115
4.3	Indices of Industrial Production, 1939–44	116
5.1	Foreign Trade Volume and Terms of Trade, 1944–50	129
5.2	Direction of Foreign Trade, 1938–50	129
6.1	Growth and Origin of National Product, 1939–80	144
6.2	Balance and Structure of Foreign Trade, 1950–60	152
6.3	Direction of Foreign Trade, 1950–60	152
7.1	Patterns of Population Growth, 1900–83	159
7.2	Distribution of Labour Force, 1948–83	160
7.3	Comparison of Official and Alton's Rates of Growth, 1961–83	162

List of Tables and Maps vii

7.4	Sources of Non-Agricultural Growth, 1953–74	163
7.5	Capital Investment and Accumulation, 1949–80	165
7.6	Structure of Industrial Production, 1939–83	166
7.7	Index of Agricultural Output per capita, 1932–81	170
7.8	Annual Average Output per capita of Agricultural Products, 1939–83	172
8.1	Structure of Major Imports, 1955–83	180
8.2	Structure of Major Exports, 1955–83	180
8.3	Foreign Trade Balance and Terms of Trade, 1961–83	182
8.4	Direction of Foreign Trade, 1960–83	188
8.5	Indicators of Living Standard, 1956–83	194/195
9.1	Cultivated Area of Personal Plots, 1965–80	211
9.2	Sources of Economic Investment, 1956–81	217
Map 1	Territory and Terrain	12
Map 2	Economic Resources	157

PREFACE

Almost twenty years have passed since I first drove into Sofia on a hazy hot August day to take up my duties as a young economics officer at the American Embassy. I returned in 1972 for postdoctoral research on Bulgarian financial institutions in the pre-1914 period. These two experiences have made Bulgaria's present and past the object of special interest in a subsequent academic career whose focus has been Balkan economic history. The editor's invitation to write the Bulgarian volume in this timely series on the economic history of the modern European states is therefore most welcome.

The series rightly recognises that the forty years which have now passed since the end of the Second World War are a long enough period to be viewed as economic history and deserve to be connected with the earlier decades of this century. This Bulgarian volume bears a special monographic responsibility. Native scholars, like their colleagues elsewhere in Eastern Europe, have — at least until recently — painted the periods before and after the Communist accession to power as contrasting so sharply that no significant links could be discerned. Western scholars have paid little attention to the earlier period, and less to relating it to the post-war era. In the present volume, I seek to spell out both the ways in which the two periods are connected and the ways in which they are not.

This volume also bears a special responsibility as a survey. The general history of modern Bulgaria, like that of most small European countries, has been neglected in the English language. The few monographs and the one excellent survey, by Richard Crampton of the period 1878–1918, listed in the bibliography are virtually all that has been published recently. I therefore feel obliged to provide the reader, at the start of each chapter, with more political background than would be required for familiar subjects such as France or the USSR.

The following treatment of Bulgarian economic history is grounded on only a few assumptions: free markets and modern technology are good, monopoly or isolation are bad, and economic value and prices are best measured by relative scarcity. Neither public nor private enterprise is presumed to be innately superior to the other.

I must draw on a variety of sources. The official publications of the Bulgarian statistical service have generally been the Balkan states' most comprehensive and careful studies throughout the twentieth century. There are none the less important gaps in these data. Native scholarship is of help here. Attention to economic history and analysis began with the founding of the Bulgarian Economic Society in 1895. Its journal was published continuously until 1940, and to a high standard. Post-war publications by a new generation of Marxist economic historians have surpassed pre-war scholarship in quantity, in detail and in use of primary sources. At the same time, this new work has concentrated on the pre-1914 period and left the years since 1948 largely uncovered. Western, and especially American, economic and statistical analysis of the post-war period has improved markedly since about 1970 and proved most valuable to this volume. Also valuable, however, have been current Bulgarian economics journals, and the publications of the Bulgarian Chamber of Commerce and Industry, the United Nation's Economic Commission for Europe, and the Council for Mutual Economic Assistance.

This volume provides me with an opportunity to update, expand and revise the treatment of Bulgaria's pre-1950 experience in *Balkan Economic History, 1550–1950: From Imperial Borderlands to Developing Nations*, which I wrote with Marvin R. Jackson for Indiana University Press (1982). Reviews of that first volume in Western and in Bulgarian journals helped to point the way towards preparing the present one. A research grant in 1981 and a travel grant in 1984 from the International Research and Exchanges Board enabled me to gather more material in Bulgaria, particularly on the post-1950 period. The Center for Bulgaristica in Sofia hosted my trip in 1984 and graciously arranged interviews with a number of Bulgarian economists and officials on recent trends. Conversations with scholars from the Higher Economics Institute 'Karl Marx' and the co-operation of Bulgarian Chamber of Commerce and Industry were especially helpful. The National Library 'Kiril and Methodius' and the library of the Bulgarian Academy of Sciences were as usual of great assistance. In the United States, the Library of Congress in Washington, DC, the National Agricultural Library in Beltsville, Maryland, and the University of Illinois Library at Urbana-Champaign proved to be useful sources of material. I consulted the latter as an Associate in the Summer Research Laboratory of the university's Russian and East European Center in 1982 and 1984.

The manuscript was read entirely or in part by Alan S. Milward of the University of Manchester Institute of Science and Technology, Marvin R. Jackson of Arizona State University, Richard J. Crampton of the University of Kent at Canterbury, and Liuben Berov of the Karl Marx Higher Economics Institute in Sofia. Their critical suggestions were helpful, although full responsibility for the final text is naturally my own. Iris Mendels provided expert editorial assistance while proofreading the final text.

The University of Maryland furnished me with indispensable assistance. A semester grant from the Graduate Research Board allowed me to draft the largest part of the manuscript. The text was then prepared on word-processor by the staff of the Department of History, principally by Darlene King. Department Chairman Emory G. Evans supported the project consistently and authorised funding for maps and statistical tables. As with *Balkan Economic History*, maps were prepared by Bowring Cartographic Research and Design, Arlington, Virginia, and tables were typed by Carol Warrington of the University of Maryland's Computer Science Center.

Again, I owe a final debt to the unfailing encouragement of my wife, Dr Anita B. Baker-Lampe, who shares my own appreciation of Bulgarian hospitality. The volume is dedicated to continued scholarly exchange between East and West, and to the proposition that scholars need not agree in order to communicate usefully or to gain from each other's work.

John R. Lampe
University of Maryland

EDITOR'S INTRODUCTION

By comparison with the nineteenth century, the twentieth has been very much more turbulent, both economically and politically. Two world wars and a great depression are sufficient to substantiate this claim without invoking the problems of more recent times. Yet despite these setbacks, Europe's economic performance in the present century has been very much better than anything recorded in the historical past, thanks largely to the super-boom conditions following the post-Second World War reconstruction period. Thus in the period 1946–75, or 1950–73, the annual increase in total European GNP per capita was 4.8 per cent and 4.5 per cent respectively, as against a compound rate of just under 1 per cent in the nineteenth century (1800–1913) and the same during the troubled years between 1913 and 1950. As Bairoch points out, within a generation or so European per-capita income rose slightly more than in the previous 150 years (1947–75 by 250 per cent, 1800–1948 by 225 per cent) and, on rough estimates for the half century before 1800, by about as much as in the preceding two centuries.[1]

The dynamic growth and relative stability of the 1950s and 1960s may, however, belie the natural order of things, as the events of the later 1970s and early 1980s demonstrate. Certainly it would seem unlikely that the European economy, or the world economy for that matter, will see a lasting return to the relatively stable conditions of the nineteenth century. No doubt the experience of the present century can easily lead to an exaggerated idea about the stability of the previous one. Nevertheless, one may justifiably claim that for much of the nineteenth century there was a degree of harmony in the economic development of the major powers and between the metropolitan economies and the periphery, which has been noticeably absent since 1914. Indeed, one of the reasons for the apparent success of the gold standard post-1870, despite the aura of stability it allegedly shed, was the absence of serious external disturbances and imbalance in development among the major participating powers. As Triffin writes, 'the residual harmonization of national monetary and credit policies depended far less on *ex post* corrective action, requiring an extreme flexibility, downward as well as upward, of national price and wage levels, than on an *ex ante* avoidance of substantial disparities in cost competitiveness and the monetary policies that would allow them to develop'.[2]

Whatever the reasons for the absence of serious economic and political conflict, the fact remains that until 1914 international development and political relations, though subject to strains of a minor nature from time to time, were never exposed to internal and external shocks of the magnitude experienced in the twentieth century. Not surprisingly, therefore, the First World War rudely shattered the liberal tranquillity of the later nineteenth and early twentieth centuries. At the time few people realised that it was going to be a lengthy war and, even more important, fewer still had any conception of the enormous impact it would have on economic and social relationships. Moreover, there was a general feeling, readily accepted in establishment circles, that following the period of hostilities it would be possible to resume where one had left off — in short, to re-create the conditions of the pre-war era.

For obvious reasons this was clearly an impossible task, though for nearly a decade statesmen strove to get back to what they regarded as 'normalcy', or the natural order of things. In itself this was one of the profound mistakes of the first post-war decade, since it should have been clear, even at that time, that the war and post-war clearing-up operations had undermined Europe's former equipoise and sapped her strength to a point where the economic system had become very sensitive to external shocks. The map of Europe had been rewritten under the political settlements following the war, and this further weakened the economic viability of the continent and left a dangerous political vacuum in its wake. Moreover, it was not only in the economic sphere that Europe's strength had been reduced; in political and social terms, the European continent was seriously weakened and many countries in the early post-war years were in a state of social ferment and upheaval.[3]

Generally speaking, Europe's economic and political fragility was ignored in the 1920s, probably more out of ignorance than intent. In their efforts to resurrect the pre-war system, statesmen believed they were providing a viable solution to the problems of the day, and the fact that Europe shared in the prosperity of the later 1920s seemed to vindicate their judgement. But the post-war problems — war debts, external imbalances, currency issues, structural distortions and the like — defied solutions along traditional lines. The most notable of these was the attempt to restore a semblance of the gold standard in the belief that it had been responsible for the former stability. The upshot was a set of haphazard and inconsistent currency-stabilisation policies that took no account of the changes in relative costs and prices among countries since 1914. Consequently, despite the apparent prosperity of the latter half of the decade, Europe remained in a state of unstable equilibrium, and therefore vulnerable to any external shocks. The collapse of US foreign lending from the middle of 1928, and the subsequent downturn of the American economy a year later exposed the weaknesses of the European economy. The structural supports were too weak to withstand violent shocks and so the edifice disintegrated.

That the years 1929–33 experienced one of the worst depressions and financial crises in history is not altogether surprising, given the convergence of many unfavourable forces at that point in time. Moreover, the fact that a cyclical downturn occurred against the backdrop of structural disequilibrium only served to exacerbate the problem, while the inherent weakness of certain financial institutions in Europe and the United States led to extreme instability. The intensity of the crisis varied a great deal, but few countries, apart from the USSR, were unaffected. The action of governments tended to aggravate rather than ease the situation. Such policies included expenditure cuts, monetary contraction, the abandonment of the gold standard and protective measures designed to insulate domestic economies from external events. In effect these policies, although sometimes affording temporary relief to hard-pressed countries, in the end led to income destruction rather than income creation. When recovery finally set in in the winter of 1932/3, it owed little to policy contributions, though subsequently some Western governments did attempt more ambitious programmes of stimulation, while many of the poorer Eastern European countries adopted autarchic policies in an effort to push forward industrialisation. Apart from some notable exceptions, Germany and Sweden in particular, recovery from the slump, especially in terms of employment generation, was slow and patchy, and even at the peak of the upswing in 1937 many countries were still operating below their resource capacity. A combination of weak real growth forces and structural imbalances in development would no doubt have ensured a continuation of resource under utilisation, had not rearmament and the outbreak of war served to close the gap.

Thus, by the eve of the Second World War, Europe as a whole was in a much weaker state economically than it had been in 1914, with her shares of world income and trade notably reduced. Worse still, she emerged from the Second World War in 1945 in a more prostrate condition than in 1918, with output levels well down on those of the pre-war period. In terms of the loss of life, physical destruction and decline in living standards, Europe's position was much worse than after the First World War. On the other hand, recovery from wartime destruction was stronger and more secure than in the previous case. In part, this can be attributed to the fact that in the reconstruction phase of the later 1940s, some of the mistakes and blunders of the earlier experience were avoided. Inflation, for example, was contained more readily between 1939 and 1945, and the violent inflations of the early 1920s were not for the most part repeated after the Second World War. With the exception of Berlin, the map of Europe was divided much more cleanly and neatly than after 1918. Though it resulted in two ideological power blocs, the East and the West, it did nevertheless dispose of the power vacuum in Central/East Europe, which had been a source of friction and contention in the inter-war years. Moreover, the fact that each bloc was dominated or backed by a wealthy and rival super-power meant that support was forthcoming for the satellite countries. The vanquished powers were not, with the exception of East Germany, burdened by unreasonable exactions, which had been the cause of so much bitterness and squabbling during the 1920s. Finally, governments no longer hankered after the 'halcyon' pre-war days, not surprisingly given the rugged conditions of the 1930s. This time it was to be planning for the future which occupied their attention, and which found expression in the commitment to maintain full employment and all that entailed in terms of growth and stability, together with a conscious desire to build upon the earlier social welfare foundations. In wider perspective, the new initiatives found positive expression in terms of a readiness to co-operate internationally, particularly in trade and monetary matters. The liberal American aid programme for the West in the later 1940s was a concrete manifestation of this new approach.

Thus despite the enormity of the reconstruction task facing Europe at the end of the war, the recovery effort, after some initial difficulties, was both strong and sustained, and by the early 1950s Europe had reached a point where it could look to the future with some confidence. During the next two decades or so, virtually every European country, in keeping with the buoyant conditions in the world economy as a whole, expanded very much more rapidly than in the past. This was the super-growth phase, during which Europe regained a large part of the relative losses incurred between 1914 and 1945. The Eastern bloc countries forged ahead the most rapidly under their planned regimes, whereas the Western democracies achieved their success under mixed-enterprise systems with varying degrees of market freedom. In both cases, the state played a far more important role than hitherto, and neither system could be said to be without its problems. The planning mechanism in Eastern Europe never functioned as smoothly as originally anticipated by its proponents, and in due course most of the socialist countries were forced to make modifications to their systems of control. Similarly, the semi-market systems of the West did not always produce the right results, so that governments were obliged to intervene to an increasing extent. One of the major problems encountered by the demand-managed economies of the West was that of trying to achieve a series of basically incompatible objectives simultaneously — namely, full employment, price stability, growth and stability, and external equilibrium. Given the limited policy weapons available to governments, this proved an impossible task to accomplish in most cases, although West Germany managed to achieve the seemingly impossible for much of the period.

Although these incompatible objectives proved elusive *in toto*, there was, throughout most of the period to the early 1970s, little cause for serious alarm. It is true that there were minor lapses from full employment; fluctuations still occurred, but they were very moderate and took the form of growth cycles; some countries experienced periodic balance of payments problems, though prices generally rose continuously, albeit at fairly modest annual rates. But such lapses could readily be accommodated, even with the limited policy choices, within an economic system that was growing rapidly. And there was some consolation from the fact that the planned socialist economies were not immune from some of these problems, especially later on in the period. By the later 1960s, despite some warning signs that conditions might be deteriorating, it seemed that Europe had entered a phase of perpetual prosperity not dissimilar to the one the Americans had conceived in the 1920s. Unfortunately, as in the earlier case, this illusion was to be rudely shattered in the first half of the 1970s. The super-growth phase of the post-war period culminated in the somewhat feverish and speculative boom of 1972–3. By the following year, the growth trend had been reversed, the old business cycle had reappeared and most countries were experiencing inflation at higher rates than at any time in the past half century. From that time onwards, according to Samuel Brittan, 'everything seems to have gone sour and we have had slower growth, rising unemployment, faster inflation, creeping trade restrictions and all the symptoms of stagflation'.[4] In fact, compared with the relatively placid and successful decades of the 1950s and 1960s, the later 1970s and early 1980s have been extremely turbulent, reminiscent in some respects of the inter-war years.

It should, of course, be stressed that by comparison with the inter-war years, or even with the nineteenth century, economic growth has been quite respectable since the sharp boom and contraction in the first half of the 1970s. It only appears poor in relation to the rapid growth between 1950 and 1973, and the question arises as to whether this period should be regarded as somewhat abnormal, with the shift to a lower-growth profile in the 1970s being the inevitable consequence of long-term forces involving some reversal of the special growth-promoting factors of the previous decades. In effect this would imply some weakening of real-growth forces in the 1970s, which was aggravated by specific factors, for example, energy crises and policy variables.

The most disturbing feature of this later period was not simply that growth slowed down, but that it became more erratic, with longer recessionary periods involving absolute contractions in output, and that it was accompanied by mounting unemployment and high inflation. Traditional Keynesian demand-management policies were unable to cope with these problems and, in an effort to deal with them, particularly inflation, governments resorted to ultradefensive policies and monetary control. These were not very successful either, since the need for social and political compromise in policy-making meant that they were not applied rigorously enough to eradicate inflation, yet at the same time their influence was sufficiently strong to dampen the rate of growth, thereby exacerbating unemployment. In other words, economic management is faced with an awkward policy dilemma in the prevailing situation of high unemployment and rapid inflation. Policy action to deal with either one tends to make the other worse, while the constraint of the political consensus produces an uneasy compromise in an effort to 'minimise macroeconomic misery'.[5] Rostow has neatly summarised the constraints involved in this context: 'Taxes, public expenditure, interest rates, and the supply of money are not determined antiseptically by men free to move economies along a Phillips curve to an optimum trade-off between the rate of unemployment and the rate of inflation. Fiscal and monetary policy are, inevitably, living parts of the democratic political process.'[6]

Whether the current problems of contemporary Western capitalism or the difficulties associated with the planning mechanisms of the socialist countries of Eastern Europe are amenable to solutions remains to be seen. It is not, for the most part, the purpose of the volumes in this series to speculate about the future. The series is designed to provide clear and balanced surveys of the economic development and problems of individual European countries from the end of the First World War to the present, against the background of the general economic and political trends of the time. Though most European countries have shared a common experience for much of the period, it is nonetheless true that there has been considerable variation among countries in the rate of development and the manner in which they have sought to regulate and control their economies. The problems encountered have also varied widely, in part reflecting disparities in levels of development. Although most European countries had, by the end of the First World War, achieved some industrialisation and made the initial breakthrough into modern economic growth, nevertheless there existed a wide gulf between the richer and poorer nations. At the beginning of the period, the most advanced region was north-west Europe including Scandinavia, and as one moved east and south so the level of per-capita income relative to the European average declined. In some cases, notably Bulgaria, Yugoslavia and Portugal, income levels were barely half the European average. The gap has narrowed over time, but the general pattern remains basically the same. Between 1913 and 1973, most of the poorer countries in the east and south (apart from Spain) raised their real per-capita income levels relative to the European average, with most of the improvement taking place after 1950. Even so, by 1973 most of them, with the exception of Czechoslovakia, still fell below the European average, ranging from 9–15 per cent in the case of the USSR, Hungary, Greece, Bulgaria and Poland, to as much as 35–45 per cent for Spain, Portugal, Romania and Yugoslavia.

Italy and Ireland also recorded per-capita income levels some way below the European average.[7]

Despite its relatively small size — a population of around 9 million today — Bulgaria deserves more attention than it has so far been accorded by contemporary historians. It has been one of the great success stories of the twentieth century, with the highest rate of economic growth in Europe and a degree of structural change second to none. Yet up to the end of the First World War, and even beyond, conditions were not especially propitious for such rapid transformation. Bulgaria had not long gained full independence and is still smarting from the effects of Ottoman hegemony. The country was very poor and not over-endowed with natural resources, while the bulk of the population derived their livelihood from the land, the development of which was impeded by geographical and climatic limitations. Non-agrarian progress before the war was confined to a narrow sector, principally in and around Sofia, a capital which increasingly tended to dominate the country. Finally, in the First World War Bulgaria backed the wrong horse and emerged in a weakened state with a reparations bill to boot.

However, it did have two distinct advantages over many other states that arose from the post-war peace settlements. First, it did not have any new territories to assimilate, nor did it incur territorial or resources losses. Second, it was not burdened with the problem of ethnic integration. Most of the population were real Bulgarians; the only important minority group was the Turks, whose numbers were dwindling. In addition, it was an advantage that its key leaders were trained in economics and perceived the need for economic progress. The problem, however, was how to capitalise on its primarily agrarian base by converting produce into a marketable surplus in order to provide the wherewithal to advance on the manufacturing front. Given the low level of efficiency in agriculture and the restricted state of world markets for primary products, there was no easy solution, short of adopting the Soviet method or the Danish approach, but the first was rejected and the second not successfully adopted. Hence Bulgaria had to rely on its own efforts to adapt the product structure of its primary base and on the flow of Western capital. Limits were thus set to the rate of transformation, and the share of manufacturing in total output remained very small, despite impressive growth in this sector during the 1920s.

The insecurity of the economic base became evident in the ensuing depression, with the steep fall in agricultural prices, the closing of Western markets and the drying up of foreign aid. Drastic steps were required to maintain the limited progress so far achieved, and these were sought in the rise of étatism, import substitution and increasing isolation through bilateral trade channels with reorientation towards the German axis in the latter part of the 1930s. Industrial growth was in fact quite spectacular at 4.8 per cent a year between 1929 and 1938, one of the highest in Europe, with import substitution accounting for much of it. Nevertheless, the development of modern mechanised manufacturing was still very limited, and if anything there was a tendency for pre-modern forms of enterprise to gain ground in this period. Sofia's importance as an economic centre became even more pronounced in these years. State control and direction of agriculture, trade and finance were more extensive than in the case of industry, despite some increase in the state-owned share in the latter sector.

In view of the closer relations with Germany during the 1930s, it was almost inevitable that Bulgaria would again select the wrong side in the forthcoming war. The consequences were more far-reaching than in the case of the earlier conflict, in that the country succumbed to Communist rule from 1944 onwards, which subsequently meant a reorientation of its economy towards planning Soviet-style, a shift in trading relationships towards the Eastern bloc, and of course the collectivisation of agriculture and the expropriation of industry. The late 1940s formed a transitional period in which the new structural format was being put into place, although it was not fully complete until later, to be followed by the first three Five-Year Plans, up to 1960. The main emphasis of the early plans was on extensive growth, whereby large amounts of inputs were channelled into a few sectors (mainly heavy industry) at the expense of the rest of the economy and regardless of efficiency criteria. Yet one cannot deny that in quantitative terms the results were highly impressive. Industry grew at a high double-figure rate throughout the 1950s, and increased its share of net product from less than a quarter to nearly one-half, with a corresponding fall in the relative size of the agrarian sector. By the early 1960s, Bulgaria had firmly established the basis of modern economic growth and structural change and had completed the shift to institutions and structure based on the Soviet planning model. The next step therefore was for a change in the nature of the growth criteria and some reform of the structure.

Subsequent plans therefore emphasised intensive growth, that is, greater attention to productive resource use as opposed to growth based on accumulating inputs. In accordance with this objective, the economic reforms of the recent past have been designed to streamline the state planning structure, decentralise ministerial control and provide greater economic incentive and initiative at the local level of operations. In part, the changes were prompted by the course of economic events, in particular, a growing labour shortage, the increasing importance of foreign trade and some slackening in economic performance after the rapid expansion of the 1950s and early 1960s.

Even so, growth in more recent years has been impressive by any standard, and Bulgaria was not troubled unduly by the oil shocks of the 1970s. Whether the economic reforms were instrumental in helping to maintain the momentum is debatable, since capital productivity improvement has been disappointing. One of the main problems in this regard seems to have been a failure of management and technology to keep pace with the hectic rate of expansion and structural change. As in other Eastern bloc countries, consumers have not secured benefits commensurate with the growth in recorded output, a reflection in part of the relative neglect of light industry, while infrastructure and environmental facilities including housing still fall well short of Western standards.

Professor Lampe's lucid account demonstrates the way in which Bulgaria has achieved its remarkable transformation from a backward agrarian economy into a highly industrialised state. He also compares and contrasts the pre- and the post-Communist periods, noting the sharp differences between the two periods, but at the same time drawing attention to the distinct elements of continuity in the country's history. Problems remain of course for the future, though perhaps these are not as acute as those facing many other countries in the Eastern sector. The fact that Bulgaria's rapid development in the Communist era has been accompanied by great political stability with relatively little internal friction or disruptive interference by the Soviet Union has undoubtedly been an important factor in facilitating the country's leap into the twentieth century, and it is possible that it may continue to be an asset in the future.

Notes

1. P. Bairoch, 'Europe's Gross National Product, 1800–1975', *Journal of European Economic History*, vol.5 no. 2 (Fall 1976), pp. 298–9.
2. R. Triffin, *Our International Monetary System: Yesterday, Today and Tomorrow* (New York, 1968), p. 14; see also D.H. Aldcroft, *From Versailles to Wall Street, 1919–1929* (1977), pp. 162–4. Some of the costs of the gold standard system may, however, have been borne by the countries of the periphery, for example, Latin America.
3. See P.N. Stearns, *European Society in Upheaval* (1967).
4. *Financial Times*, 14 February 1980.
5. J.O.N. Perkins, *The Macroeconomic Mix to Stop Stagflation* (1980).
6. W.W. Rostow, *Getting From Here to There* (1979).
7. See Bairoch, 'Europe's Gross National Product', pp. 297, 307.

Map 1: Territory and Terrain

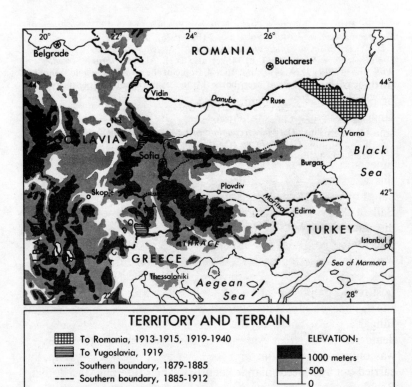

INTRODUCTION

The image of Bulgaria in Western eyes has remained uncertain throughout the twentieth century. Less attention, whether scholarly or popular, has been paid to Bulgaria in the English or French languages than to any other Balkan state except Albania.[1] This small country's political history has helped to turn the somewhat cloudy Western perception dark on more than one occasion. 'Three times in my life,' observed Winston Churchill acidly in 1944, 'has this wretched Bulgaria subjected a peasant population to all the pangs of war and chastisements of defeat.' The loss of the Second Balkan War in 1913 had indeed inaugurated the bitter, unavailing dispute over Macedonia with the neighbouring Yugoslavs, and led directly to fatal Bulgarian alliances with Germany in the two world wars. Few Western observers, Churchill included, have bothered to look carefully into these misfortunes, let alone to search for positive accomplishment.

The national catastrophes, as they became known within Bulgaria, contributed to the inter-war failure of parliamentary democracy. The great Agrarian leader, Aleksandur Stamboliiski, was only the first victim in a series of assassinations, principally carried out by Macedonian irredentists. Their Internal Macedonian Revolutionary Organisation became synonymous with political terrorism throughout Europe. Within the Bulgarian government, first party and then royal, non-party regimes contributed to growing political intolerance. The immediate post-war period brought still more uncompromising and ruthless consolidation of political power, now under Communist auspices.

Since then, co-operation with the Soviet Union has been so close, particularly in foreign policy, that many Western observers have found it difficult to discern a separate Bulgarian identity. But there is one, and its complexity should not be obscured by the enduring Soviet alliance or the small size of the country, barely 9 million people. A remarkable economic history has contributed both to this separate identity and to close relations with the USSR.

Bulgaria is a relatively new state, autonomous since 1878 and fully independent only since 1908. In part because of its youth, the country has been in the forefront of two significant economic trends in twentieth-century Europe. Within a spectrum of generally rapid growth for national product and exports across the continent, Bulgarian rates of growth have been the highest of all. At the same time, scarcities of labour and capital have held back these advances periodically.

A further, more political trend is hard to quantify, but no less profound a change. It is the rise of state economic control and initiative. For Bulgaria, this process began long before the Communist accession to power in 1944. It has now evolved into unusually persistent efforts to reform a centrally planned, Soviet-style economy in the face of the declining rates of growth and increasingly scarce resources which confront all European countries.

According to the rough calculations of Paul Bairoch, Bulgaria's gross national product grew by an average of 4.4 per cent a year across the entire period 1913–73. This rate was the highest among all European countries, whose average growth was 3 per cent. Only the USSR came close to Bulgaria, with 3.9 per cent.[2] For 1913–50, Bulgaria's average increase of 2.7 per cent trailed only Sweden and Norway, and almost doubled the European average of 1.4 per cent. This occurred despite an overwhelmingly agricultural economy. For the period 1950–73, the Bulgarian growth rate of 7.2 per cent still exceeded the European average of 5.6 per cent, but less dramatically. Only the doubtful Romanian figure is higher.

More reliable figures for foreign trade point to another area of Bulgarian pre-eminence. The small Balkan state increased its share of European export value more rapidly before and since the First World War than any other country. That share tripled between 1890 and 1910 to reach 0.3 per cent and then quadrupled between 1910 and 1970 to touch 1.2 per cent.[3] Only the Scandinavian countries have approached these increases. By the early 1960s, Bulgarian trade turnover, the sum of exports and imports, had passed its inter-war peak, 38 per cent of national income and by 1984 reached 100 per cent, obviously growing more rapidly than the economy as a whole. This disproportion suggests export-led growth. By the early 1970s, the value of Bulgarian exports and imports per capita equalled not only the European average, but also the Yugoslav and Romanian values combined. Thus the relatively small share of Bulgarian foreign trade conducted with Western Europe in recent decades, barely 15 per cent, bulks larger when considered in absolute terms.

The Western role in Bulgaria's large foreign trade, if we exclude undivided pre-war Germany, has all the same been surprisingly small throughout the twentieth century. Even during the heyday of Bulgarian wheat exports before the First World War, the Western European share of Bulgarian trade turnover did not amount to 30 per cent. It had already fallen to its present proportion by the 1920s. Then German purchases of tobacco and various foodstuffs pushed their share of Bulgarian exports close to 40 per cent by 1929. Thus the stage was set for still larger German predominance during the depression decade of the 1930s, now conducted according to bilateral clearing agreements. These cut the Bulgarians off from international prices and the market mechanism, such as they were in the 1930s. Bulgaria's trade since the Second World War has also been based on similar bilateral agreements, although on more favourable terms, with the Soviet Union and other Eastern European states. Trade with the USSR has consistently accounted for at least half of Bulgarian turnover since 1945. The continent's most rapidly growing export sector has therefore been more closely tied to a single buyer for a longer time than any other European country.

The origins of state control and initiative in Bulgarian economic history predate the clearing agreements with Nazi Germany. Barred from instituting protective tariffs under the terms of its pre-1908 ties to the Ottoman Empire, the young government had in 1894 promulgated the first formal industrial legislation in the Balkans. By 1903, the region's first state bank for agricultural credit had opened in Sofia. The First World War and its aftermath occasioned a series of state controls over agricultural marketing that spread from exports to domestic sales. Their renewal and expansion after 1930 preceded the clearing agreements with Germany and continued into the Second World War. So did state encouragement and control of industry, if not much investment or ownership. What sort of groundwork was laid for the transition to collectivised agriculture and centrally planned state industry after 1945, and also for subsequent efforts to reform them? How has a small economy so long oriented to foreign trade adapted to a system which had first evolved in the huge isolated economy of the inter-war Soviet Union?

The post-war period has none the less proved pivotal. It has witnessed the structural shift of labour and capital into modern industry, which typically turns growth into development.

16 Introduction

Industry's pre-war shortage of both inputs abated, at least until the 1970s. Manufactured goods, including foodstuffs, now account for the great majority of exports. Five of the nine chapters that follow deal with these forty years since the Second World War.

Geographic Limitations

The physical geography of Bulgaria is not particularly favourable to modern economic development.[4] The wooded mountain ranges that dominate western Bulgaria and extend eastward across its centre are poor in mineral resources, apart from low-grade coal and small non-ferrous deposits (see Maps 1 and 2). Petroleum deposits are smaller still, making the modern Bulgarian economy the most dependent on energy imports (over 70 per cent of consumption) in Eastern Europe.

Agricultural land is also less than ideal. The soil of much of the western uplands is suitable only for tobacco or vegetable cultivation. Further east, grain cultivation on the rolling plain to the north of the Balkan mountains faces limited rainfall. The climate of south-eastern Europe is generally less congenial to crops in this regard than that of the north-western agricultural plain. An ample supply of moisture from the Atlantic has been reduced by mountain ranges and the continental landmass to less than 750 mm a year by the time the flow of weather systems from the west reaches the Bulgarian lands. The most fertile soils in the north-east and south receive less than 500 mm. Periodic droughts and irregular rainfall are therefore inevitable.

The overwhelming size and isolated location of its capital city has also distinguished the small Bulgarian economy. The combination is an unusual one in European economic history. Sofia counted over 150,000 residents in a national population of 4.8 million in 1920, some 3 per cent, but by 1980 had over 1.1 million inhabitants, of a total of 8.9 million, or 13 per cent. Bulgaria's present density of 81 people per km^2, already low by European standards, is thus lower still, if Sofia is excluded. Throughout the century the capital has been the centre not only of political power, but also of modern manufacturing and of urban culture and consumption. Making the city's political role more important was the general absence elsewhere in the country of any strong regional interest or ethnic separatism. The mixture of interests and peoples in neighbouring Yugoslavia that has led logically to a federal form of government never existed in Bulgaria.

The present population of 9 million is over 85 per cent ethnically Bulgarian. That proportion has hardly changed since the Second World War, despite the departure of 90 per cent of the country's 60,000 Jews for Israel and some 200,000 Turks for Turkey. Turks had begun leaving when the present borders were essentially fixed in 1913, after the Second Balkan War.[5] They and the other large minority, the Gypsies, have remained politically powerless, although their combined numbers now exceed one million.[6] All of this has allowed central control to flow more smoothly from Sofia than from any of the other Balkan capitals, probably facilitating the post-war adoption of Soviet-style planning. The sister city in Eastern Europe would appear to be Budapest, especially in the recent decades of reforming the same centralised Soviet system. The Hungarian capital has nearly 20 per cent of an equally small, even more homogeneous population of 10 million.

Unlike most other large European capitals, including Budapest, Sofia is not, however, a port city. Its location on an upland plateau in the mountainous western quarter of the country places it close to the large deposits of lignite and lesser ones of hard coal and copper that are Bulgaria's principal mineral resources (see Map 2). Yet the city is far from the country's only navigable river, the Danube, which forms the northern border with Romania, and farther still from the Black Sea. In addition, Sofia has been less of a crossroads for Western visitors and ideas than for Central or Eastern European influences. The isolation of this large capital city from Western Europe and within the country itself may help us to understand the strength of the state's role in Bulgarian economic growth throughout the twentieth century.

Notes

1. For evidence of this neglect in Western economic analysis, see John R. Lampe, 'The Study of Southeast European Economies: 1966–1977', *Balkanistica*, vol. IV (1977/8), pp. 63–88.

2. Paul Bairoch, 'Europe's Gross National Product, 1800–1975', *Journal of European Economic History*, vol. 5, no. 2 (1976), table 17, p. 305.

3. Paul Bairoch, 'European Foreign Trade in the XIX Century', *Journal of European Economic History*, vol. 2, no. 1 (1973), table 5, p. 17.

4. For a technical review of Bulgaria's physical geography in regional perspective, see Roy. E.H. Mellor, *Eastern Europe, A Geography of the Comecon Countries* (New York: Columbia University Press, 1975), pp. 3–34, 306–13.

18 *Introduction*

5. At that time, the 96,345 km² that had comprised Bulgarian territory since 1886 increased to 111,836 km². The expansion occurred at Ottoman expense, mainly along the western two-thirds of the southern border with what now became an enlarged Greek state. The inter-war loss of the southern Dobrudja to Romania, and a small reduction of the western border in favour of the new Yugoslav state trimmed Bulgaria's territory to 103,146 km². The former's restitution in 1940 established the post-1944 dimensions of 110,669 km².

6. Some 340,000 Turks emigrated from Bulgaria between 1912 and 1939, joined by another 200,000 since 1944. Their share of the total population thereby fell from 11.6 per cent in 1910 to 10.2 per cent in 1934 to 8.6 per cent by 1956. High birth rates for these mainly rural Turks pushed their share back to 9.1 per cent by 1965, the last year that an ethnic breakdown was officially published. Despite the departure of another 41,000 since then, the 1965 total of 747,000 Turkish-speakers had probably grown past 900,000 by 1980. The 197,000 Gypsies recorded in the 1956 census may now exceed 350,000. Bulgarian Muslims or Pomaks, remain about 160,000. See Marvin R. Jackson, 'Changes in the Ethnic Content of the Balkan National Populations, 1912–1970', *Faculty Working Papers*, Department of Economics, EC 83/84−20, Arizona State University, forthcoming in *Southeastern Europe*; and also Radio Free Europe Research (RFER), *Bulgarian Situation Report*, 2, 30 January 1985.

1 INITIAL GROWTH, 1878–1918

At the start of the nineteenth century the Bulgarian lands were still under closer Ottoman control than the Sultan's other Balkan territories. So they had remained since the Ottoman Turks had overwhelmed Bulgarian feudal forces over four hundred years before. From the 1820s forward, however, a remarkable cultural revival spread through the predominantly Bulgarian population of the upland towns. Primary schools and adult reading rooms became the crucial institutions. The result was a rediscovery of Bulgarian national identity, virtually lost since the Ottoman conquest, except in isolated monasteries.[1]

Political independence came less quickly. Until it did, in 1878, we cannot speak of a national economy. By 1870, only approval for the religious autonomy of the Bulgarian Orthodox Church had been won. Ottoman authorities struck down a nascent revolutionary movement with death sentences. They brutally suppressed the wider insurrection of April 1876. It remained for tsarist Russia, flushed with Pan-Slavic sympathy for its 'little brothers', to drive Ottoman forces from the territory of what became the Bulgarian state in 1878.

The political prospects of the new state seemed bright at first. An autonomous Bulgarian government had come into being by 1879, under a liberal constitution which favoured the legislative branch. The next forty years were not kind to those early prospects. British and German influence had vetoed the initial Russian settlement of the Russo-Ottoman war of 1877–8. The Treaty of San Stefano had created a large Bulgarian state stretching to the Aegean Sea. The subsequent Treaty of Berlin not only barred any Russian prince from the throne, but denied the new state the Bulgarian lands south of the Balkan mountains until 1885 and prevented the incorporation of Macedonia entirely (see Map 1). Yet the presence of Russian advisers until 1885 strengthened the ministries at the expense of the *Subranie*, or legislative assembly.[2] After Tsar Alexander III of Russia forced young Alexander of Battenberg to abdicate, a second German prince, Ferdinand of Coburg, began his long rule in 1886. He capitalised on the assassination in 1895 of the principal

Bulgarian parliamentary leader, Stefan Stambulov, to undertake a process of dividing the emerging liberal and conservative political parties. By the outbreak of the First Balkan War in 1912, 'Foxy Ferdinand' had split these parties into nine quarrelling factions and had himself named tsar.[3] His ministries ran the country from Sofia. His one personal achievement was to arrange full Bulgarian independence from the Ottoman Empire in 1908; his greatest failure the vainglorious effort, less for reasons of state than of religious fantasy, to capture Istanbul from the Ottoman Empire during the First Balkan War in 1912.

The leadership of the large Bulgarian army and ministerial bureaucracy, and all of the political parties except the Agrarians and Narrow Socialists must share the blame with Ferdinand for the two 'national catastrophes' that followed in 1913 and 1918. One was the Second Balkan War, fought against Bulgaria's former allies in the first. Defeat left Macedonia still in Serbian hands and cost the small Bulgarian state access to the Aegean Sea and the grain-exporting southern Dobrudja. Bulgaria entered the First World War in 1915 on the side of the Central Powers in a futile effort to reverse these territorial dispositions. Final defeat in 1918 forced Ferdinand into exile and ended his unfortunate influence.

The troubled and eventually tragic political history summarised above should not obscure the more consistent record of economic and educational advance which characterised this long period, particularly the last pre-war decades. The century preceding the disastrous end to the First World War witnessed enough agricultural growth to triple domestic commerce and to increase foreign trade at an even faster pace. Monetisation and modern commercial practice spread into the countryside from upland towns. It was from there that Bulgarian artisan manufacture expanded during the mid-nineteenth century. Mechanised factories mushroomed after 1900. This was still small-scale, private production. Some state initiatives to promote modern growth also emerged from the government ministries and two semi-official banks in Sofia. Although not successful in promulgating a single, coherent set of policies, these public institutions had all the same established a clear predominance over the private centres of economic power by the First World War.

The Ottoman Economic Legacy

The long Bulgarian experience as an integral part of the Ottoman Empire was complex. Native scholars reject the old picture of unrelieved repression and backwardness,[4] but the eighteenth century and the first decades of the nineteenth remain difficult to paint in lighter colours. Disease, disorder and corruption descended. Local Muslim warlords, mainly Turks, carved out their own fiefdoms in the lowland valleys. Trade routes became unsafe, tax collection arbitrary and excessive. Record-keeping deteriorated too much to leave any useful statistical account of the period. All of this plainly stood in the way of economic growth or modernisation.

The land regime in the rural Bulgarian lowlands of the early nineteenth century remained, moreover, less fair and less efficient than the original Ottoman system. At its sixteenth-century apex, the Sultan's Muslim cavalry officers, or *sipahi*, had collected low and fixed proportions of peasant grain harvests without even the right to pass on the duties of tax collection to their sons. By the start of the eighteenth century a new system of private, inheritable landholding had replaced the original one. Primarily Turkish officers and officals used this new *chiflik* system to keep control of the Bulgarian lowlands well into the mid-nineteenth century. They exacted payment for use of good pasture lands and forced Bulgarian peasants who wished to cultivate arable land into sharecropping agreements which exacted half of the grain harvest or more. Ottoman authorities in Constantinople had already despaired of collecting their share of such exactions in the eighteenth century and began selecting ethnic Bulgarian traders, village elders or other notables, the *chorbadzhii*, to be their new tax collectors.

These impositions did not create a series of large, consolidated estates practising capitalist agriculture on the Polish or Prussian pattern. The Bulgarian *chiflik* were too small and unconsolidated for that.[5] They were, none the less, exploitative. As a result, the Bulgarian peasantry moved away from the *chiflik* estates of the Muslim notables and from the taxes of the native *chorbadzhii* by deserting the lowlands in general. They retreated into the less accessible uplands of the Balkan mountains, or migrated across the Danube into the Romanian principalities, especially after the renewed disorders attending the Russo-Ottoman War of 1828–9.

Yet a restoration of public order and economic activity began in the 1820s. Another half century of Ottoman domination followed, but would provide three stimuli to the formation of a modern economy. The first came precisely from the Balkan uplands and the proto-industrial development there of artisan manufacture.[6] What in parts of early modern Western Europe had been seasonal artisan production became almost year-round activity in upland towns like Gabrovo and Karlovo after 1830. Peasants who had migrated from the lowlands did not return for the summer harvest, except where *chiflik* land could be had for money rentals, as in the Plovdiv area. The impetus that put this large labour supply to work, manufacturing primarily woollen cloth and braid, plus some shoes and ironware, was growing demand from the wider Ottoman market. The needs of the Sultan's new standing army, created in 1826, soon resulted in numerous military contracts. Some included the army's delivery of raw wool or other inputs for local manufacture under what the Ottomans called the *ishleme* system. (In Chapter 5, similar arrangements to use raw cotton from the USSR before and after the Second World War are described.) Improved public order also made trade routes more secure. The Ottoman reform decree of 1839 gave Bulgarian merchants the formal right to trade freely throughout the empire. A large colony of them had already been doing business in Constantinople, a city of half a million people. Artisans followed. By 1870, this expanding commercial network supported a dozen towns with populations over 5,000 and a labour force of perhaps 100,000 upland artisans, nearly 10 per cent of the active labour in the Bulgarian lands. Bills of exchange supplied short-term credit and compensated for the inability of the Ottoman economy to accept paper money.

What happened, in brief, was the spread of monetisation (the money-goods relationship in Marx's terms) and the creation of a broad base of commercial experience in the Bulgarian lands. The profits and travel of successful artisan owners and traders encouraged them to pay for building local schools and for the further education of their sons outside the Ottoman Empire, in Russia or the German lands. Eventually, however, this large sector of small-scale, unmechanised manufacture would create a barrier to the introduction of modern factory production that endured into the inter-war period.

The second Ottoman stimulus also promoted monetisation and market experience. Under Western European pressure, the Porte had lifted its long-standing ban on the export of wheat from the empire in 1842. The Bulgarian lands responded more quickly than any other Ottoman territory to this access to the expanding European market. By 1850 Bulgarian exports of wheat and barley, half to the hungry British market, had risen sharply. Varna's trade turnover had moved ahead of Thessaloniki's to become the largest of any Balkan port. Then came several bad harvests in the 1850s. Combined with lost rights of tax collection, they prompted a few Turkish *chiflik* holders to sell off all or part of their lands. The Ottoman land code of 1858 guaranteed property rights to the new Bulgarian owners. They appeared in increasing numbers during the 1860s, as Bulgarian grain exports doubled in value. Turkish sales of land had become significant by the early 1870s. By then the population of the Bulgarian lands had risen to almost 3 million, double the 1830 estimate. The Muslim, largely Turkish proportion remained about 30 per cent of the total. Ethnic Bulgarian movement from upland to lowland was at the same time large enough to create a severe labour shortage for artisan manufacturers. Wages rose, prompting some limited mechanisation, as we shall see.

The third Ottoman stimulus was the central government's effort after 1860 to introduce modernising economic reform in return for increased tax collection. Under European pressure to repay a rapidly accumulating sum of foreign loans, Ottoman authorities could hardly fail to notice the large export surpluses accumulating in the Bulgarian lands. Their nothern region was therefore selected in 1864 as the site for a single enlarged province, to be run on principles of efficiency and modern improvements.[7] An able Pomak, or Bulgarian Muslim, Midhat Pasha, was put in charge of this new Danubian region, or *vilayet*, with Sofia as his headquarters. Despite its location and relatively small size, the city thus became a leading candidate for the capital of a future Bulgarian state. Midhat's attempts to begin construction of modern roads and to open a network of co-operative savings banks for agriculture were important more as precedents for the future than for what they actually accomplished. He did succeed in raising the collection of tax revenues, mainly levied on crops and livestock, by one-third in three years. Three-quarters were then spent outside the Bulgarian lands. From this, Bulgarian landholders could only conclude that they would be better off to break away from the Ottoman Empire, as the revolutionary movement was now urging.

Grain Exports and Peasant Agriculture after 1878

For the Bulgarian peasantry, the immediate consequence of the break in 1878 with the internal regime of the Ottoman Empire was the hasty departure of perhaps one-half of the remaining Muslim population. Enough had returned by 1880 to bring Turkish-speaking numbers back to 700,000.[8] This was 25 per cent of the estimated population in the new state and the Rumelian south, incorporated in 1885 (see Map 1). Turkish officers, officials and also smallholding peasants scrambled to sell off their holdings at reduced prices throughout the decade. Turkish speakers were 132,000 fewer by 1892. (The Treaty of Berlin had revoked the outright seizures which Bulgarian peasants had carried out in the wake of the Russian army's victory.) These properties were often larger and used more as pasture than the average Bulgarian holding, but were soon subdivided for sale to individual peasant households. The new supply of better valley land left its new owners with what one Bulgarian scholar has estimated to be over 40 million leva (or francs) in debt, typically owed to Bulgarian merchants or officials who had made the initial purchase.[9] This burden, plus obligations remaining from land purchases in the 1870s or earlier, meant that a large fraction of the Bulgarian peasantry had acquired smallholdings which would have to generate a marketable surplus for debt repayment or be sold off again.

The growth of Bulgarian grain cultivation and the exports that followed until the Balkan wars of 1912–13 seem at first glance to have been an 'engine of growth', in the manner sometimes ascribed to Canadian grain during the same period. The Bulgarian national product appears by very rough reckoning to have kept pace with growth rates in the developed European economies, rather than falling behind as Paul Bairoch has argued. More clearly, its value per capita by 1911 amounted to one-quarter of the German figure, far beyond the tiny fractions of the developed economies' average recorded by most of today's Third World countries.[10] Agriculture still accounted for about three-quarters of gross Bulgarian production, and more of the net, even at the end of the period.

Grain hectarage and output per rural inhabitant would continue to increase even after 1900, when those indicators had begun to decline in neighbouring Romania and Serbia. Exports had risen even faster, outstripping continued population growth. Bulgaria still relied on grain for 70 per cent of export value in 1911, barely

Table 1.1: Gross Social Product in 1911

	Million leva	Leva per capita	Real growth rate per capita, 1904–11 (%)[a]
Crop agriculture	611	139	0.5
Livestock	217	48	
Forestry	97	22	
Large-scale private industry	112	26	13
Small-scale private industry	78	18	
Mining	1	0.3	
Other, including trade	310	71	

[a] Population growth averaged 1.5 per cent a year for 1904–11.

Source: J.R. Lampe and M.R. Jackson, *Balkan Economic History, 1550–1950* (Bloomington, Indiana: Indiana University Press, 1982), tables 6.1 and 6.2, pp. 162–3.

Table 1.2: Real Exports per capita, 1886–1910 (annual average)

	Million leva	Per-capita value (leva)	Real per-capita value (1906–10 prices)
1886–90	62	19.7	28.4
1891–5	78	23.2	31.6
1896–1900	69	19.0	24.5
1901–5	120	30.8	40.3
1906–10	130	30.9	30.9

Source: Lampe and Jackson, *Balkan Economic History, 1550–1950*, table 6.5, p. 168.

less than the 72 per cent of 1886–90.

Closer inspection of foreign trade and agricultural production, however, reveals a less promising picture. Bulgaria recorded a significant export surplus, some 25 per cent, only for the period 1901–5. Totals mainly of grain exports to Western Europe were otherwise matched by totals mainly of manufactured imports from Germany and Austria-Hungary. Rising international grain prices after 1900 put Bulgaria's *real* per-capita exports for 1906–10 back to the level of 1891–5, as Table 1.2 indicates.

Behind these falling real values lay the top-heavy structure of Bulgarian exports. They were overly dependent on wheat and corn sales. The swelling American, Canadian and Argentinian export of these two grains to Western Europe threatened to reverse the post-

Table 1.3: Distribution of Cultivated Land, 1897 and 1911 (percentages)

	Grain	Industrial crops	Vegetables	Vineyards, fruits	Forage crops
1897a	74.0	0.3	2.4	5.0	18.0
1911b	75.9	1.3	3.4	2.6	16.8

a Total of 2.45 million hectares.
b Total of 3.32 million hectares.
Source: K. Popoff, La Bulgarie économique, 1879–1911 (Paris: 1920), p. 145.

1900 rise in international prices, even if the First World War had not intervened. The war and its aftermath were to do worse to Bulgarian sales (see Chapter 2).

Another danger from over-dependence on one or two staple exports is of course vulnerability to bad harvests. Grain crops are particularly dependent on sufficient and regular rainfall. Bulgaria's most fertile soils in the north-east and south receive less than 500 mm of rainfall a year, as noted in the Introduction. The confluence of droughts and irregular rainfall in the period 1897–1900 was, however, the worst in modern Bulgarian history. By this time cereal grains covered 76 per cent of cultivated land and 23 per cent of total arable land. No significant change had occurred by 1911.

Two-thirds of the grain produced was winter wheat, 30 per cent of which was exported. Three failed harvests out of four at the end of the century cut wheat production in half and total export value by one-third. Good weather for 1901–5 allowed metric tonnage harvested to rise by one-half, but the average figure slipped again by 20 per cent for the next five-year period, because of bad harvests in 1907 and 1909. Only rising wheat prices, up 40 per cent for 1906–10 over their 1895 low point, allowed wheat sales to maintain their share of export value.[11]

Peasant Reaction and Protest after 1900

By this time the Bulgarian peasantry had taken matters into their own hands. While augmenting the wheat and other grain land under cultivation by one-third since 1900, they did not increase marketed sales nearly so fast. The tonnage of wheat exported actually fell by 25 per cent for 1906–10 versus 1896–1905. Corn exports, mainly from the larger, share-cropped holdings of the north, rose by just 5 per cent between these two periods and were still just three-quarters of the total value for wheat. Smallholders on properties under 20 hectares owned over 75 per cent of private agricultural land. They reacted economically in two ways.

One was to divert some of their wheat and corn crops, barley and oats as well, from the market-place to household consumption. This tactic not only allowed them to protect their standard of living from another shock of the sort that it had received before 1900. It also allowed them to increase their numbers of pigs per capita by 23 per cent and sheep by 7 per cent between 1900 and 1910.[12] These animals provided more household necessities than the cattle and horse population, which declined. Exports of livestock or processed meat meanwhile stayed the smallest of any of the Balkan states, less than 2 per cent of export value, and draft animals were in short supply.

Another reaction was to develop different sorts of primary export. The tenfold rise in egg sales, mainly to Germany, from 1891–9 to 1907–11, allowed the value of animal products to maintain its 10 per cent share of exports. The other agricultural exports to increase were also unprocessed and required no refrigeration or other modern packing. Turnips, kidney beans, and silk cocoons together accounted for another 7 per cent of 1911 exports. This variety showed the readiness of the Bulgarian peasantry to respond to market signals.

It did not prevent the stagnation of per-capita exports after 1905 and their decline in real terms. Despite the limited movement into new markets, we cannot escape the reduced real growth of grain exports. The smallholding peasantry was largely free from sharecropping or plantation-like restrictions. It rationally chose to feed and provide for its growing numbers. The Bulgarian population had passed 4.3 million by 1910, from 2.8 million in 1880, and was still 81 per cent rural. Natural increase continued to rise at 1.5 per cent a year during the last pre-war decade. Sustaining this rise was rural consumption at or above subsistence, as well as the lowest rural density among the independent Balkan states, 45 persons per hectare.

The prospect for long-term economic growth in such circumstances, however, was poor. Agriculture outside the market-place or outside further value added by the manufacturing process, even if over the level of subsistence, has never encouraged modern

Table 1.4: Ownership of Private Rural Land, 1897 and 1908 (percentages)

	Under 3 ha	3–5 ha	5–10 ha	10–20 ha	20–50 ha	Over 50 ha
1897a	11.3	11.0	26.6	26.4	14.7	10.0
1908b	11.6	10.8	26.8	27.8	14.5	8,7

a Private rural land totalled 3.98 million of 7.42 million occupied hectares, the latter covering 76.9 per cent of the state's territory.
b Private rural land totalled 4.63 million of 7.98 million occupied hectares, the latter covering 82.9 per cent of the state's territory.

Source: Popoff, *La Bulgarie économique, 1879–1911*, pp. 87–97.

economic development in Europe or elsewhere. Where agriculture has played the leading role, as with Denmark's dairy production before the First World War, a marketed surplus of increasingly processed exports has been responsible. Even a larger Bulgarian wheat surplus would not necessarily have prompted a boom in flour milling, given the small size of the local urban market and the predominance of Budapest and Bucharest in regional exports.

The great majority of the Bulgarian peasantry continued to concentrate on wheat and corn cultivation during the last pre-war decade, doubtless encouraged by rising international grain prices and generally good harvests. Some of the smallholdings were simply too small to afford a worthwhile marketed surplus. Table 1.4 reveals that one-third of the country's private owners held less than 2 hectares, and another 30 per cent 2–5 hectares. Literally no trend toward larger holdings appeared between the 1897 and 1908 data. Worse still, these holdings were unconsolidated, typically divided among ten or more scattered plots. Half of the holdings under 2 hectares were in fact plots away from the peasant's own village, so-called *parakende* land.

Modern agricultural equipment and infrastructure was also wanting. A British consular report of 1893 blamed the first deficiency on the smallholders' illiteracy and resistance to new tools or techniques. Between 1893 and 1910, however, the number of iron ploughs and other small implements grew almost tenfold. Only 18 per cent of all peasant households had an iron plough as a result, but the proportion was reportedly 35–40 per cent in the main grain-growing areas. The use of steam-powered machinery stayed very small, despite a sixfold rise in reaping and threshing units imported from 1900 to 1911. The few relatively large landowners used up to 40 per cent of this machinery. Yet they typically rented out a large part of their property to share-croppers and furnished them with no equipment. These share-croppers, typically smallholders supplementing their income, accounted for most of the hired rural labour, whose numbers almost tripled between 1900 and 1910. By then they were only 9 per cent of the active total in agriculture. Hired labour remained in short supply. Little wonder that wheat and corn yields showed no pattern of significant increase after 1900.[13]

According to the staple theory of pre-1914 growth, rising grain exports were supposed to attract investment capital for railways, port facilities and other infrastructure. For reasons that the next section will make clear, this kind of capital did not come from Western Europe to Bulgaria in the proportions that went to Canada or even neighbouring Romania, with its huge share-cropping estates and greater mechanisation. Regular tax revenues paid for fully half of the railway and port construction in Bulgaria from 1888 to 1912. Railway connections from the main grain-growing areas to the Danubian ports of Ruse and Svishtov and to the Black Sea outlets of Burgas and Varna were only partly complete after 442 km of new northern lines opened in 1899. Modern port facilities at Burgas and Varna were also slow to appear. Large ships could not dock at the latter until 1906. The inland storage silos and elevators that were needed to collect smallholders' sales were not built at all.

Peasant smallholders trying to cultivate grain or other crops more intensively for the market were discouraged by both the tax structure and the supply of credit. The imposition of money taxes, first on harvested crop value in 1882 and instead on agricultural property in 1894, had undoubtedly pushed peasants further into the market-place to pay what they owed, especially in a period of falling grain prices. The Bulgarian government of the late 1890s sought to repair growing budget deficits by revising the tax structure once again. It decided to replace the land tax with a harvest tithe in kind, just as grain prices were turning upward. The proposal coincided with the series of disastrous harvests noted above. Peasant protests led by the newly formed Bulgarian Agrarian National Union (BZNS) forced its abandonment.

The government followed the general European fiscal trend for the rest of the pre-war period, and turned to indirect taxes. By 1911 they accounted for twice the revenue from direct taxes, nearly reversing the relationship of 1900. This policy had the effect, according to one British scholar, of reducing the effective tax burden of the peasantry in a time of rising prices.[14] They would normally consume less than urban residents of the manufactures that bore high sales taxes by 1910. They also faced similar taxes on some of the goods they sold in the market-place. In return, direct state investment in agriculture consisted of little more than importing the mulberry trees that had allowed silk cocoons to rise to 1–2 per cent of export value.

Limited agricultural credit also discouraged peasant smallholders from consolidating their holdings and from farming them more intensively. Access to mortgage loans and especially to short-term credit undoubtedly improved after 1903, under the joint auspices of the new Bulgarian Agricultural Bank and the rural credit cooperatives established by the Agrarian Union. It was nevertheless insufficient.

The influence of the BZNS had declined after its victory against tithes in kind. Some 400 hastily organised producers' co-operatives soon disintegrated. Party fortunes revived after 1903, under the charismatic leadership of Aleksandur Stamboliiski. Trained as a schoolteacher, like the other Agrarian leaders, he was a powerful public speaker and a tireless organiser.[15] His desire to eliminate the monarchy and to organise a Bulgarian republic according to occupational 'estates', with the peasantry predominant over all urban interest groups, made him a dangerous radical to Ferdinand's successive governments. These prevented the Agrarians from winning more than 15 per cent of the vote in any pre-war election. They could not, however, prevent Stamboliiski's Agrarians from rebuilding the network of rural co-operatives, now primarily as credit unions on the Raiffeisen model, but sometimes with insurance, storage and agricultural extension services as well. Their numbers jumped from 68 in 1904 to 1,123 by 1908. Some 576 cooperatives with 40,000 members banded together to join the new Central Co-operative Bank in 1910.

The state-supervised Bulgarian Agricultural Bank (Bulgarska Zemedelska Banka) had helped the co-operatives to establish this new institution and had already extended them crucial lines of current-account credit. Zemedelska Banka was itself a consolidation into a single bank in Sofia of the rural savings banks introduced under Ottoman rule and revived by the new national government in 1880. Opening in 1903, the consolidated bank's assets grew to one-third of the Bulgarian total by 1911 (see Table 1.6). Mortgage loans, drawn on livestock as well as landholdings, amounted to 15 per cent of the Zemedelska's assets by then. Short-term credit in sums from 500 to 1,000 leva (equivalent to French francs) were lent to peasants directly or through the BZNS cooperatives. Large by Balkan standards, these credits allowed peasants to purchase seed, fertilisers or iron ploughs. The bank also promoted the introduction of one new crop, sugar beet now refined as we shall see for domestic consumption, not for export. These short-term loans did not, however, result in significant purchase of steam-powered equipment for grain cultivation, nor did they promote any noticeable consolidation of scattered peasant plots.

In other words, the best system of agricultural credit in the pre-1914 Balkan states was not good enough to spread the more varied and intensive cultivation mandated by the grain crisis of 1897–1900. Had financial conditions permitted, the rising exports of eggs, kidney beans, turnips and silk cocoons might have increased enough to compensate for the relative decline in wheat sales. Their marketing might also have been taken over by the cooperative network from the few European trading firms which had exclusive rights to their export.

State Finance within the European Monetary System

The rest of Bulgaria's financial structure was a major restriction facing the relatively extensive network for agricultural credit which emerged during the last pre-war decade. Despite a late start in the 1880s, it was already in place well before the Zemedelska Banka was consolidated. That structure unfortunately lacked the strongest feature of the Western European model, active commercial banks, and relied instead on the weakest feature, a state central bank committed to restricted note issue, government loans and an overvalued exchange rate. In return for this commitment, the Bulgarian government was nevertheless to receive less favourable consideration from the Western European capital market, based in Paris, than any other Balkan state.

The initial Bulgarska Narodna Banka had opened unsuccessfully in Sofia in 1879. All of its modest paid-in capital had been contributed by the state treasury. State funds constituted all of the 10 million leva (francs) worth of paid-in capital, when the original institution was successfully reconstituted as a joint-stock bank of issue. It reopened for business in 1886, further fortified by a 10 million leva loan from Germany's Deutsche Bank. Prince Ferdinand reserved the right to appoint the bank's entire administrative council. Thus wedded to the state, the reorganised bank had established the silver-backed leva as the country's only legal tender within a few years.[16] Unlike the Ottoman Empire, Bulgaria's commercial framework was therefore sufficiently modernised to accept domestic paper money by the 1890s.

The subsequent pre-war history of the Bulgarska Narodna Banka unfortunately revolved around continuing state control and newfound aspirations to the gold standard. The obvious attractions of conversions to the gold-backed leva were twofold. It would eliminate the gold premium for international use of silver-backed notes and also bring the government more and cheaper access to the European capital market.

At first, everything went wrong with the Bulgarian plans. The government's effective borrowing on the European bond market had totalled just 202 million francs by 1896, mainly for construction of the aforementioned Orient Express line across Bulgaria and for payment on the Ottoman Public Debt. The effective interest rate was high — 6.6 per cent, or one point more than that paid for much larger sums by Balkan neighbours.[17] By this time the government owed nothing to the Ottoman Public Debt, whose administration was primarily in French hands. In 1896, after some brief interest from the American Rockefellers, the ownership of the Oriental Railway Company had passed into German hands, those of the Deutsche Bank. Its attempt to undercut freight rates on the Bulgarian state lines soon led to a major dispute. Sofia's reaction was to start construction parallel to the Orient line east of Plovdiv, but leading north to the Bulgarian port of Burgas, rather than south to Ottoman territory and Constantinople, as did the existing line. French representatives on the Debt Administration blocked Bulgarian efforts to borrow funds to finish the parallel line or to buy out the Deutsche Bank. They joined forces with German interests to deny the Bulgarian government any access to European capital on acceptable terms until 1902.

To make matters worse, the agricultural crisis of 1897–1900 cut into the government's anticipated revenues and made repayment even of existing debt difficult. Only a series of expensive short-term loans on the European money market covered the deficits. The

Table 1.5: Main Financial Indicators, 1886–1911 (in million leva)

	State budget revenues	Exports	BNB emission	Foreign debt
1886	49	37	0.1	
1890	69	70	2.0	
1895	77	84	19.4	
1900	81	96	18.2	355
1905	128	125	37.2	
1910	181	198	81.6	
1911	191	181	110.8	589

Source: Lampe and Jackson, *Balkan Economic History, 1550–1950*, tables 7.2, 7.3, 7.7 and 7.8, pp. 212, 219, 233–4.

Table 1.6: Distribution of Bank Assets in 1911 (in franc equivalents per capita)

	Central bank	Other state banks	Private native banks	Foreign banks	Total
Bulgaria	71	54	23	21	167
Romania	109	66	73	42	290
Serbia	72	57	110	24	263
Greece	176	0	45	134	355

Source: Lampe and Jackson, *Balkan Economic History, 1550–1950*, table 7.4, p. 223.

burden of these years on state revenue and central-bank note issue was as severe, according to Table 1.5, as the upward turn was sharp during the subsequent decade.

The Bulgarian contribution to the upturn consisted of more than enjoying better weather and accepting Franco-German terms. Budgetary reforms, similar to those in the other Balkan governments after the Greek bankruptcy of 1897, introduced European standards of accounting. Revenue forecasts became more reliable, and the state budget was more predictably balanced. New indirect taxes and more efficient state enterprise (see the next section) doubled annual revenue between 1902 and 1911 to 199 million leva. European lenders, now two-thirds French, accepted Bulgarian state bonds in the effective sum of 464 million francs at only 5.5 per cent interest between 1902 and 1911.

The price of these funds, however, was high in other respects. The 85 million franc French loan eliminated the expensive floating debt from short-term borrowing, but tied repayment to state revenues from the tobacco monopoly. The large loans of 1904,

1907 and 1909 were secured by claims on indirect tax or customs revenues, and thereby increased state reliance on them. The Bulgarian government's use of the new loans was itself questionable. Only 21 per cent was now directed to railway and harbour construction or other economic purposes, compared with 37 per cent of the earlier loans. The rest was divided almost equally between foreign-debt service, the expenses of an expanding state bureaucracy, and the increased military spending that followed the bloody Ottoman suppression of the Macedonian uprising in 1903.

The commercial banking system paid perhaps the greatest price exacted by renewed access to the European capital market. The central bank strengthened its position still further. Unlike its Balkan counterparts, however, it continued to lend heavily to the government. Such lending quadrupled during the last pre-war decade to reach one-quarter of the bank's assets. This large fraction mitigates the stimulus to the private sector of a long-overdue increase in note issue, recorded in Table 1.5. The increase could have been much larger, had the bank not kept its reserve ratio over 50 per cent in order to protect the leva's parity with the gold franc, finally achieved by 1906. According to Table 1.6, Bulgaria's private commercial bank assets, foreign affiliates included, remained the smallest in the independent Balkan states on a per-capita basis.

What held back the growth of native banks was not only restricted note emission and central bank discounting before 1905, but also the continuing predominance of the two state banks, the Narodna and the Zemedelska, especially in Sofia. Apart from a branch of the small Bulgarska Turgovska Banka of Ruse, no rival institution opened its doors in the capital city until 1905. Afterwards the new banks were foreign-owned, but affiliates, rather than the branches of major European institutions with the much larger access to loanable funds that they would afford. The combined founding capital of the German Kreditna Banka, the largely French Generalna and the largely Austrian Balkanska Banka amounted to 10 million leva, only one-half of the enlarged central-bank figure. Their long-term activities eschewed any of the direct investment in infrastructure postulated by the staple theory. Their assets were instead concentrated in municipal loans and in the state tobacco and match monopolies. Their equally cautious short-term lending was restricted to international trade. The reduction in range of interest rates, which typically signals the integration of a nationally competitive banking system, did not occur. Bulgarian banks still charged the highest rates of any Balkan financial structure in 1911, despite the relatively low 8 per cent levied by the Agricultural Bank.[18]

A European-style financial structure had indeed been created in barely twenty years. Yet it remained too top heavy, given the size of the two state banks, to provide even short-term credit in large enough quantity and low enough rates for expanding agricultural exports or import-substituting manufacture. The overvalued leva which this structure did assure only added to the demand for credit, making exports more expensive abroad and imported manufactures cheaper at home.

Modern Manufacturing and its Limits

The top-heavy structures of the Bulgarian financial system and export market did not bode well for the growth of modern industry. Nevertheless, its annual growth during the last pre-war decade was the most rapid of any of the Balkan states. The real per-capita increment for large-scale manufacturing between 1904 and 1911 was 13 per cent, compared with 10 per cent for Serbia and 5.3 per cent for Romania during 1901–11.[19] This impressive rate of growth forms the strongest part of the argument by Bulgarian Marxist historians, from their founding father and party leader Dimitur Blagoev forward, that an industrial revolution was already under way in pre-war Bulgaria.

Simply citing the very small base from which this growth began does not seem a sufficient answer to the Marxist argument. Had growth continued at the pre-1914 pace for the next thirty years, a Bulgarian economy unquestionably close to converting the majority of its labour and capital to industrial activity would have existed by the time that a Communist government took power in 1944. A more persuasive response would begin by noting that over half of the 1904–11 growth occurred in the last two years, hardly a permanent trend. Further counter-argument comes from problems of supply and demand which would, even in the absence of the two world wars, have cut modern industrial growth back from its minispurt of the last pre-war decade.

The supply of capital for self-sustained growth was not promising. The fixed capital already invested in mechanised

Table 1.7: Industrial Output, Capital and Firm Size in 1911

	Gross output (million leva)	(Net) (million leva)	Fixed capital (million leva)	Average no. of workers per firm	Average HP per worker
Textiles	21.6	(9.5)	15.9	67	0.92
Flour	46.3	(5.0)	15.9	11	5.64
Other foodstuffs	18.6	(10.6)	18.5	24	2.32
Leather	5.5	(0.9)	2.8	18	1.27
Chemicals	4.2	(2.5)	3.3	17	1.05
Paper	0.9	(0.4)	1.6	51	1.53
Wood processing	2.7	(1.2)	3.3	63	0.61
Metal processing	3.0	(1.5)	3.3	56	0.42
Construction materials	4.7	(2.9)	6.4	116	0.57
Total	107.3	(34.5)	82.1	39	1.72

Source: Lampe and Jackson, *Balkan Economic History, 1550–1950*, tables 8.2 and 8.3, pp. 242–3.

manufacture by 1911, as recorded in Table 1.7, was only slightly larger per capita than the Serbian figure and just one-third of the Romanian figure. Average horsepower per worker was under 1.5, or half the Romanian averages, for all branches except flour and foodstuffs. The variety of specialised production remained very limited. Gross output per capita in 1911 was still one-quarter less than the small Serbian output.[20] Even if war had not intervened, such an industrial sector could have looked forward neither to economies of scale nor the wide spread of European best practice. Lack of specialisation would in any case have discouraged sales outside the limited domestic market.

Where could Bulgarian industrial firms hope to find the investment funds and the short-term credit needed to expand fixed capital beyond the very small foundation of 1911? Less than 15 per cent had incorporated themselves. The chances for public sale of stock issues were in any case limited. Dividends from incorporated manufacturing amounted to only 8 per cent, compared with the earnings of 10 per cent or more reported by joint-stock banks and trading companies. Retained earnings were not therefore a likely source of new capital.

The lower profit available in industry provides one obvious reason why native or foreign banks generally avoided stock ownership or other long-term investment. Native banks were further constrained, as we have seen, by their small size. The two largest even then cut back the share of their assets directly invested in industry from 10 per cent to 5 per cent during the last pre-war years. Of the five banks with European backing, by 1911, all had political origins in Great Power rivalries, which put no premium on Balkan industrial development. The banks' principal direct investments, outside the state monopolies, were in cement and sugar manufacture, plus coal and copper mining. Private industry received perhaps 3–4 million leva from them by 1911, plus another 3–4 million directly from Belgian mining interests. The combined sum was just 10 per cent of total investment in private industry.[21]

Short-term credit came no more easily. The huge Bulgarian central bank trimmed the industrial share of its loans on discounted bills of exchange and current-account overdrafts from 20 per cent down to 10 per cent during the last pre-war decade. According to the complaints of Bulgarian manufacturers, the new European-backed banks in Sofia after 1905 drained away much of this largest source of loanable funds. Remaining funds were in any case given overwhelmingly to manufacturers in the Sofia area, where joint-stock industrial profits were one-half of the modest national average. For their part, the European banks consistently charged industrial borrowers interest rates 4–5 percentage points higher than their other customers paid.

The supply of labour also limited the potential for further industrial growth. After 1900 it was scarce and increasingly expensive. Rising prices on the international grain market combined with the rapidly growing populations of the capital cities to push up urban food prices some 30–40 per cent by 1911 in all the Balkan states.[22] Labour protests and the young socialist movement helped bring wages up to cover most of these increases, if not to provide sufficient income for housing as well. Sofia's industrial wages were the highest of any of the Balkan capitals. It had after all been a commercial backwater without significant artisan industry before independence in 1878. Constructing a European-style capital and supporting state-encouraged industry had attracted some Bulgarian labour, but not enough. State employees still outnumbered industrial workers 3 to 1 in 1910. The resulting shortage and higher wages attracted migrant labour from Serbia, at least for construction.

The large force of artisan labour created in the upland towns along the Balkan range was not much attracted. Enough stayed to keep artisan numbers, part-time rural producers excluded, at half again the total employment of 16,000 in large-scale manufacturing. The rest, spinners and weavers of woollen textiles, had left for lowland agriculture long before 1900. Where 60,000 or 70,000 artisans had been employed in the upland towns during the 1860s, the new textile factories had only 1,200 employees by 1903 and 4,300 by 1911. Factory production had risen from 11 per cent to 82 per cent of Bulgarian woollen output between 1870 and 1903; it was now the largest branch of Bulgarian industry, apart from flour milling. Over the same period, according to the British economic historian Michael Palairet, so many artisans left to become peasant smallholders and to raise sheep with lower-quality wool for their own consumption that the per-capita Bulgarian consumption of manufactured wool dropped by almost one-half.[23]

This reduction in domestic woollen demand was perhaps exceeded by increases for other sorts of manufacture, but not by much. Sofia's population had risen rapidly from 20,000 in 1880 to 102,000 by 1910, but a majority of that increment had migrated from other Bulgarian towns.[24] The country's overall urban population (towns exceeding even 2,000) remained virtually unchanged at 19 per cent. The prospects for domestic Bulgarian demand before 1914 were thus modest.

In addition, imports of European manufactures now absorbed a significant share of the domestic market, foodstuffs largely excepted, as they had not done in Ottoman times or even in the pre-1900 period. Textiles were the largest single *import* throughout the pre-war decades, still amounting to 28 per cent of the total 1911 value. Their 55 million leva-worth almost tripled the 21 million leva of native, large-scale manufacture. Import substitution would have to come before export expansion for industrial growth to continue, or so it seemed to contemporary Bulgarian observers.

The Industrialisation Debate and Economic Policy

Modern industry and what to do about it lay at the heart of an ongoing debate over economic policy. The debate occupied the centre of the Bulgarian political stage for much of the pre-1914 period. The issue of industrialisation was more widely and hotly debated in Sofia than in any of the other Balkan capitals. In Athens, Bucharest and Belgrade, economic issues were secondary. Not so in Sofia. The key to national assertion seemed from the start to lie more with choosing the correct strategy for economic development than with pursuing the seemingly futile goals of parliamentary democracy or the claim to Macedonia. Prince Ferdinand would not hear of one, nor the Great Powers of the other.

An initial economic debate in the 1880s had dealt with railway construction. The questions were where to build the cross-country lines and who was to build them. By the turn of the century, some 1,345 km of largely east-west track had been added to the 221 km of the original 1868 line from Ruse to Varna. All but the Oriental line's 379 km east of Plovdiv were state-owned. Commerce had still to jump sharply upward once the new lines were open. Internal trade rose by only one-quarter and external trade even less from 1885 to 1895.[25] Something more was needed, but what?

The start of the industrialisation debate may be found in early issues of the journal of the Bulgarian Economic Society, founded in 1895. The country's agricultural crisis, which began the following year, was undoubtedly a catalyst. For solutions, educated Bulgarians drew on schools of thought in Germany or Russia, where the great majority of several thousand university graduates had received their training. More of the country's political leaders appear to have been trained specifically in economics than their counterparts elsewhere in the Balkans, where lawyers or engineers predominated.

By the turn of the century there were few leaders opposed entirely to modern industry. Their conservatism typically derived not from European ideas, but from nativist opposition to railways, European capital and modern industry as vehicles for 'foreigners from whom we have more to fear than an epidemic'. Russian populist ideas and immigrants had informed the Agrarian Union's opposition to industry, when it was first established in 1899.[26] Its later revival under Aleksandur Stamboliiski saw the BZNS favour the continued existence of artisan manufacture, but not decry modern industry, as opponents often charged, unless it resorted to cartel restrictions. The answer then, according to Stamboliiski's repeated statements in Subranie debate, was not to cripple modern industry, but rather for the state to take over control of the cartels.

By the last years a rough consensus was also emerging among the other political leaders to go ahead with industrialisation. The only stipulation was to protect artisan and agricultural interests, especially against abuses by foreign capital. Even the ardent industrialiser Ivan Estatiev Geshov, Finance Minister in the 1890s and Prime Minister for 1911–13, came to favour the promotion of food processing, despite his vision of Bulgaria as another Belgium, because of their common coal resources. He was a German-style National Liberal, in fact, if not in party label, whose ardent protectionism belied his economics training in the stronghold of free trade, Great Britain. Dimitur Blagoev, head of the 'Narrow' Socialist Party which broke with the 'Broads' in 1903, was more faithful to his schooling in Russian Marxism. He argued for industrialisation as the precondition for socialist revolution. The majority of other political figures drew instead on German stage theory and Friedrich List's protectionist doctrine to advocate the protection of infant industry, preferably using agricultural raw materials, so that the Bulgarian economy could move to a higher stage.[27] They continued to disagree over how to control foreign investment, but most agreed that it was necessary. The majority also shared the enthusiasm of Andrei Liapchev, Agriculture and Trade Minister during 1908–10 and later Prime Minister during 1926–31, for agricultural co-operatives (Stamboliiski's BZNS aside) and for the promotion of crops other than grain.

The same sort of consensus emerges on import tariffs and legislation to encourage private industry. Here were the two main policy tools available to any pre-1914 government. The terms of Bulgaria's 1882 agreement with the Ottoman Empire allowed renegotiation of the basic Ottoman rate of 8 per cent *ad valorem* only every seven years. After an 1890 agreement with Great Britain, the Bulgarian government signed treaties with all the Great Powers and with Serbia in 1897 for a general rate of 14 per cent plus a few rates specifically higher. The first comprehensive Bulgarian tariff regulation was carefully prepared between 1900 and 1905. Its entirely specific rates set maxima averaging 25 per cent *ad valorem*. That level seemed to follow the protectionist example set by Romania and praised by Geshov and his allies in the Subranie. In fact, the Romanian average was 33 per cent *ad valorem*, mainly because of a 29 per cent rate for manufactures, compared with just 20 per cent for Bulgaria.[28] The great majority of Subranie delegates favoured some increase, but were reluctant to offend agricultural interests by passing a genuinely protectionist tariff. This emerges clearly from the actual ratio of tariff revenues collected to total import values. The Bulgarian ratio was fully 21 per cent. Imports were little deterred, in contrast to Romania with an actual ratio of only 11 per cent.

Initial Growth, 1878–1918   41

Legislation for industrial encouragement in pre-1914 Central and Eastern Europe consisted of exempting private firms from certain taxes and import tariffs, as well as a reduction in railway freight rates. Geshov himself drafted the first Bulgarian legislation, passed in 1894 for ten years' duration. It applied to heavy industry, textiles and sugar refining. Revisions followed in 1905 and 1909. The 1905 provisions were to last for twenty years and were extended to food, if not tobacco processing. To qualify, firms had to employ a minimum of 20,000 leva capital, 15 employees and 5 horsepower, all slight reductions. By 1909, registered firms accounted for just over 75 per cent of production from all enterprises meeting the minima. Unregistered firms presumably wished to avoid the increasingly rigorous requirements that this legislation introduced for financial reporting and at least some tax payment.

The economic impact was, however, uncertain. Combined exemptions for 1912 amounted to 4 per cent of industrial capital, but industrial profits showed no clear upturn for encouraged firms. They still lagged behind bank and trade profits.[29] The exemptions encouraged concentration in Sofia, their political source, still further. The customs exemptions also led a number of Sofia manufacturers to import more inputs than needed in order to sell them on the open market. In 1909 encouragement was extended to firms with as few as 10 employees. Some were plainly artisan enterprises, thus subverting the original intention of promoting modern manufacture.

Other, more direct initiatives, although never co-ordinated under a coherent programme, laid more groundwork for the state control of industry than did tariffs and encouragement laws. Mineral lands, limited to coal and some copper, were largely state property. The state coal mines, principally the Pernik complex near Sofia, accounted for 90 per cent of the coal mined in Bulgaria and over 60 per cent of its consumption. Industrial firms had to rely on British imports and often complained of shortages. The state railway took the majority of domestic coal.

From 1901 to 1911, the state railway system had grown by over one-third to 1,928 km. The acquisition of the 435 km held by the Oriental Railway Company in 1908 was soon followed by the construction of 300 km of badly needed connecting lines in 1909–10.[30] The railways undoubtedly contributed much to the doubling of internal and foreign trade between 1901 and 1911. Yet the network was still not large enough, as in the Romanian case, to stimulate the growth of domestic metallurgy or machine production. The system nevertheless earned revenues in 1911 that amounted to 14 per cent of state budget receipts. Railway earnings and those of the several state monopolies had risen almost as fast as indirect taxes after 1900. Together these earnings acounted for one-third of the 1911 budget. Without these earnings, in other words, budget receipts could not have increased at 11 per cent a year during the last pre-war decade, one percentage point faster than exports.

The tobacco monopoly earned twice the revenue of the match and other monopolies. Its operation under licence to the large European-backed banks helped ensure foreign-debt repayment. Arbitrary price increases also served this end, while keeping exports to a minimum. Its 40 enterprises, from a total of 58 tobacco processors, were tied together under a cartel arrangement that state regulations had sanctioned in 1905. The arrangement merged a number of smaller firms and soon covered 70 per cent of production, by far the most comprehensive of the half-dozen pre-war cartels.[31]

The State's Economic Role in the First World War

More direct state control of the economy received a special impetus, as elsewhere in Europe, from the First World War. The two Balkan wars of 1912–13 were too brief to bring pressures to bear for mobilising the entire economy. The four wartime years from 1915 to 1918 were more than long enough. In 1915, Bulgaria's army of 800,000 men joined German and Austro-Hungarian forces in ousting the Serbian army from Macedonia. Then it settled into holding the so-called Salonika Front on Bulgaria's southern border. The army faced an ever stronger force assembled by the Entente powers. This Allied force, although buoyed by a re-equipped Serbian army, was unable to break through the Bulgarian defences of the south-eastern Macedonian uplands until September 1918. Without some success in organising the resources of the home front, the costly Bulgarian war effort could not have continued as long as it did.

For agriculture, the war years began and ended badly. The Second Balkan War of 1913 had seen the loss to Romania of the southern Dobrudja (see Map 1), a territory which had accounted for 20 per cent of grain production. Bad weather in 1914 held the harvest of all crops to 79 per cent of the 1908–12 average on post-1912 territory. By 1915, however, the pre-war level of production, Dobrudja excluded, had nearly been regained. It would not slip badly again until 1918. Then drought and a disintegrating military situation combined to keep crops to 52 per cent of the pre-war level. Even grain imports from Germany, which exceeded exports, failed to prevent near famine.

State controls contributed significantly to both initial success and ultimate failure.[32] In 1915 the government created the Committee for Public Welfare (KZOP) to set prices for authorised exports, essentially to the Central Powers, and to ban all others. It empowered the central and agricultural banks as well as the local, or *obshtina* governments to purchase all grain and milk, and to control the sale of flour mills. This apparatus worked well enough for exports, but could not control the domestic market, especially where the army's General Commission for Requisitions took an interest. The Subranie soon expanded the Committee's powers to encompass all areas of the economy related to the war effort. Its plans for cultivated acreage were enforced through the co-operative network. Reorganised in 1916 as the Central Committee for Economic and Public Welfare, it was now responsible for military as well as civilian supplies and was authorised to control production directly. Although cultivated area, agricultural labour, draft animals and transport were down by over one-fifth from pre-war levels, production held up better. The general staff continued to press for control of its own foodstuffs. In 1917, this civilian Committee was reorganised under military auspices as the so-called Directorate. Army agronomists were assigned still wider powers over crop cultivation. Troops were brought in to help with the harvest. Forced requisitions began. These measures were supposed to increase the delivered surplus and to restrain a growing black market. The Directorate's decision to keep delivery prices low only boosted the black market. By 1918, peasant resistance to these military requisitions became widespread. The food stores that the local government had opened in town now had empty shelves. The rationing coupons that the central government had introduced in 1916 became worthless.

Private industry went through a similar experience. The growing powers of the government's Committee and then the Directorate placed 42 per cent of all industrial labour and fully 60 per cent of production under state control by 1917. New cement and sugar cartels added to official leverage. So did the severalfold increase in the value of joint-stock incorporations, principally in Sofia, where the ministerial bureaucracy or the Directorate promised war contracts. Coal mines, already 90 per cent state-owned, as we have seen, doubled their production to compensate for the end of imports. The supply of labour and imported inputs had by 1918 dropped enough to cripple industrial production. Only 30 per cent of large-scale manufacturing firms still worked regularly, two-thirds of them for army contracts. Artisan shops began to go out of business.

Bulgaria's wartime alliance with Germany encouraged the increasingly military effort to mobilise all resources in both of their now isolated economies. Their sources of supply in Western Europe had been severed. The German war effort was assumed, wrongly as it turned out, to have been the most successful among the European powers. The Bulgarian alliance with Germany included copying this model for mobilisation. It extended to trade and finance, as well as to industry. Like the later alliance of 1941−4, the economic consequences were in the end as disastrous as the political ones.

The initial Bulgarian experience was promising. The government's large budget deficits of the period 1912−14 had been accompanied by import surpluses nearly as large. German loans, beginning with 270 million leva through the Berliner Disconto-Gesellschaft in July 1914, had cut that year's budget deficit in half. Subsequent German loans and subsidies would cover half of the Bulgarian deficits for the period 1914−18. Huge German purchases of Bulgarian tobacco replaced the lost Belgian grain market. Tobacco sales, which were 77 per cent of the total exports, generated trade surpluses until 1918. Over half of export value went to Germany throughout. The leva held to 75 per cent of its pre-war value until the end of 1916, before falling to 50 per cent by 1918.[33]

Less favourable features of the German alliance increasingly came to the forefront. From the start, the expensive Bulgarian military presence in Macedonia did not give its government control of the most valuable economic resources there. Mines were placed in German hands. So was the railway system, eventually extending to the Oriental Railway line from Sofia to Nis and Belgrade. German authorities took over the Bulgarian copper mine at Plakalnitsa from its previous Western European ownership, and purchased a majority share in the main coal mine at Pernik to ensure access for mining operations. Minerals, tobacco and also grain were exported to Germany at prices well below the international level. Trade agreements in 1916 and 1917 ratified these terms. German collection centres for the forced delivery of agricultural produce operated not only in Macedonia and occupied Serbia, but in Bulgaria proper. These goods, as well as unauthorised purchases or seizures by individual German soldiers, could all be taken north to the Fatherland on a railway system free of Bulgarian control. Thus the quantity and value of Bulgarian food and mineral exports to Germany recorded in Sofia for 1916–18 was one-half or less of the totals set down as received at the German end of the line.[34] This drain undoubtedly added to shortages and inflation in Bulgarian towns during the last years of the war.

The near famine which overtook a defeated Bulgaria in the fall of 1918 was only the most visible sign of an economy in crisis. Macedonia was abandoned, along with any hope of access to the Aegean Sea. A depreciated currency and a discredited government were still in place. Back from the front came a once proud army, now demoralised. It had fought for four years at a cost of 101,000 lives, after losing 54,000 in the Balkan wars, and had lost in the end. Rapid recovery in the post-war period seemed an unlikely prospect, especially if grain exports or convertible currency on the pre-war pattern would be required for loans from the victorious Entente powers. Only the arrival of an emergency purchase of American wheat would prevent literal starvation in 1919.

Notes

1. A good brief summary of the Bulgarian national revival may be found in Barbara Jelavich, *History of the Balkans*, vol. I (Cambridge: Cambridge University Press, 1983), pp. 335–48.
2. See Charles Jelavich, *Tsarist Russia and Balkan Nationalism: Russian Influence in the Internal Affairs of Bulgaria and Serbia, 1879–1886* (Berkeley, California: University of California Press, 1958); and George F. Kennan, *The End of Bismarck's European Order* (Princeton, New Jersey: Princeton University Press, 1979), pp. 103–219.
3. From the aptly titled biography by a descendant of the politically eminent Danev family, Stephen Constant, *Foxy Ferdinand: Tsar of Bulgaria* (New York: Franklin Watts, 1980). The one recent Western survey of the country's history throughout this period concentrates instructively on the unfortunate dynamics of internal politics. See Richard J. Crampton, *Bulgarian 1878–1918: A History* (New York: Columbia University Press, 1983).
4. See, for instance, Nikolai Todorov, *The Balkan Town, 15th–19th Centuries* (Seattle, Washington: University of Washington Press, 1983); and Nikolai Todorov

et al., Stopanska istoriia na Bulgariia, 681–1981 [Economic History of Bulgaria, 681–1981] (Sofia, 1981), pp. 162–213.

5. For a summary of the evidence that the *chiflik* covered perhaps 20 per cent of cultivated land and occupied 10 per cent of peasant labour by the early nineteenth century, see John R. Lampe and Marvin R. Jackson, *Balkan Economic History, 1550–1950: From Imperial Borderlands to Developing Nations* (Bloomington, Indiana: Indiana University Press, 1982), pp. 133–7.

6. Ibid., pp. 139–45. Also see Todorov, *The Balkan Town*, chs 4–6. The model for 'proto-industrialisation', as posited by Franklin Mendels for some of eighteenth-century Western Europe, calls in part for rising artisan manufacture from seasonally unemployed agriculture labour that reaches beyond the local market. Franklin Mendels, 'Proto-industrialization: The First Phase of the Process of Industrialization', *Journal of Economic History*, vol. XXXII, no. 1 (1972), pp. 241–61.

7. Lampe and Jackson, *Balkan Economic History*, pp. 149–52. Also see Roderick H. Davidson, *Reform in the Ottoman Empire, 1856–1876* (Princeton, New Jersey: Princeton University Press, 1963), pp. 136–71.

8. Crampton, *Bulgaria, 1878–1918*, table 1, p. 176.

9. Liuben Berov's evidence for this estimate is summarised in Todorov *et al., Stopanska istoriia na Bulgariia*, pp. 217–35. Nearly one-fifth of the cultivated area had changed hands between 1878 and the 1897 land census. Crampton, *Bulgaria, 1878–1918*, pp. 177–86.

10. The estimate for average annual growth of 0.5 per cent for 1860–1910, compared with 1 per cent for 20 European countries, found in Paul Bairoch, 'Europe's Gross National Product, 1800–1975', *Journal of European Economic History*, vol. 5, no. 2 (1976), p. 283, rests on doubtful trade data for the early period, which probably understates the rate of Bulgarian growth. The best Bulgarian estimate, in Kiril Popoff, *La Bulgarie économique, 1879–1911* (Paris, 1920), p. 516, is much higher, 2 per cent a year for 1892–1911, but uses uncertain data for 1892 and a 1911 figure which is undoubtedly overstated. For further discussion, see Lampe and Jackson, *Balkan Economic History*, pp. 160–5.

11. On price movements, see Zhak Natan *et al., Ikonomika na Bulgariia*, vol. 1 (Sofia, 1969), p. 363; on grain area and output, Lampe and Jackson, *Balkan Economic History*, tables 6.7–6.8, pp. 171–2.

12. Lampe and Jackson, *Balkan Economic History*, table 9.6, p. 293. On the distribution of agricultural exports, see *Statisticheski godishnik na Bulgarskoto Tsarstvo* [Statistical Yearbook of the Bulgarian Kingdom], *1911* (Sofia, 1912), pp. 304–13.

13. Table 6.13, in Lampe and Jackson, *Balkan Economic History*, p. 188, shows wheat and corn yields of 9.3 and 10.1 quintals per hectare for 1906–10, compared with 9.9 and 10.4 for 1896–1905. British consular reports detailing Bulgarian agricultural practice may be found in British Parliamentary Papers, *Annual Series*, XCII (1893), LXXXI (1901), no. 2642, no. 1300, XCVII (1904), no. 3236, CXXII (1906), no. 3630, CIX (1908), no. 4132, and XC (1911), no. 4609. *Statisticheski godishnik na Bulgarskoto Tsarstvo 1912*, p. 165, provides a complete breakdown of agricultural equipment in use for 1893 and 1910.

14. Michael Palairet, 'Land, Labour and Industrial Progress in Bulgaria and Serbia before 1914', *Journal of European Economic History*, vol. 12, no. 1 (1983), pp. 81–96. For a contrary view on indirect taxation and detail on the struggle over tithes and the land tax, see Crampton, *Bulgaria, 1878–1918*, pp. 206–8, 260–7, 358.

15. See John W. Bell, *Peasants in Power: Alexander Stamboliski and the Bulgarian Agrarian National Union, 1899–1923* (Princeton, New Jersey: Princeton University Press, 1977), pp. 22–84.

16. The last Russian silver-backed roubles and Ottoman silver coins had by then

been driven from their final refuge, Austrian construction sites for the Oriental Railway route across Bulgaria to Constantinople. Jelavich, *Tsarist Russia and Balkan Nationalism*, pp. 65–73, 96, 107. On the founding of the Bulgarska Narodna Banka, see John R. Lampe, 'Finance and pre-1914 Industrial Stirrings in Bulgaria and Serbia', *Southeastern Europe*, vol. II, no. 1 (1975), pp. 35–42.

17. See Lampe and Jackson, *Balkan Economic History*, table 7.7, p. 233. For further details on Bulgarian borrowing, see Lampe, 'Finance', pp. 35–78; and Tsvetana Todorova, *Diplomaticheska istoriia na vunshnite zaemi na Bulgariia, 1878–1912* [A Diplomatic History of Bulgarian Foreign Loans, 1878–1912] (Sofia, 1971), pp. 11–27.

18. Table 7.5, in Lampe and Jackson, *Balkan Economic History*, p. 225, shows that private Bulgarian banks had a net profit on total assets, a surrogate for interest levels, of 2 per cent in 1911, compared with figures under 1 per cent for Romania and Serbia. On the various activities of Balkan domestic and foreign banks, see ibid., pp. 219–35.

19. See Lampe and Jackson, *Balkan Economic History*, table 6.2, p. 163. See Marvin R. Jackson and John R. Lampe, 'The Evidence of Industrial Growth in Southeastern Europe before the Second World War', *East European Quarterly*, vol. XVI, no. 4 (1983), pp. 393–3.

20. Tables 8.2–8.3, in Lampe and Jackson, *Balkan Economic History*, pp. 242–3, include Romanian and Serbian data. Also see John R. Lampe, 'Varieties of Unsuccessful Industrialization: The Balkan States before 1914', *Journal of Economic History*, vol. XXXV, no. 1 (1975), pp. 56–85.

21. Lampe and Jackson, *Balkan Economic History*, pp. 254–62; Asen Chakalov, *Formi, razmer i deinost na chuzhdiia kapital v Bulgariia, 1878–1944* [Structure, Extent and Activity of Foreign Capital in Bulgaria, 1878–1944] (Sofia, 1962), pp. 54–8.

22. See Lampe and Jackson, *Balkan Economic History*, table 8.4, p. 245; also pp. 238–44.

23. For an extensive case-study of Bulgaria and Serbia, see Michael Palairet, 'The Decline of the Old Balkan Woollen Industries, 1870–1914', *Vierteljahrschrift für Sozial und Wirtschaftsgeschichte*, vol. 70, no. 3 (June, 1983), pp. 331–62.

24. The growth of Sofia is treated in John R. Lampe, 'Modernization and Social Structure. The Case of the pre-1914 Balkan Capitals', *Southeastern Europe*, vol. 5, pt 2 (1979), pp. 11–32.

25. Natan *et al.*, *Ikonomika na Bulgariia*, vol. I, pp. 348–50.

26. Bell, *Peasants in Power*, pp. 16–21.

27. Zhak Natan (ed.), *Istoriia na ikonomicheska misul v Bulgariia* [A History of Economic Thought in Bulgaria], vol. II (Sofia, 1973), pp. 70–8, 156–81, 275–88.

28. See Lampe and Jackson, *Balkan Economic History*, table 8.8, p. 266. Liapchev himself gives a thorough review of Bulgarian tariff policy and sets out his own advocacy of agriculture-related protectionism in *Spisanie na bulgarskoto ikonomichesko druzhestvo*, vol. VIII, no. 10 (1904), pp. 697–721.

29. Lampe, 'Varieties of Unsuccessful Industrialization', pp. 79–85. On the several industrial laws, see Crampton, *Bulgaria, 1878–1918*, pp. 372–4.

30. The Russian loan of 1909 confirmed the acquisition of the Oriental Railway Company line, claimed by the Bulgarian government since its declaration of complete independence from the Ottoman Empire in 1908. On railway and port construction, see *Doklad do Ferdinand I ot Ministarski Savet, 1887–1912* [Report to Ferdinand I from the Council of Ministers, 1887–1912] (Sofia, 1912) pp. 711–35.

31. On the tobacco monopoly and tax structure, see Natan *et al.*, *Ikonomika na Bulgariia*, vol. I, pp. 384–6, 402–4; and *Statisticheski godishnik, 1912*, pp. 349–57.

32. For a brief account see Crampton, *Bulgaria, 1878–1918*, pp. 491–500;

Todorov *et al.*, *Stopanska istoriia na Bulgariia*, pp. 287–94; and, for details, G. Danailov, *Les effets de la guerre en Bulgarie* (Paris, 1932).

33. On wartime foreign trade and finance, see Crampton, *Bulgaria, 1878–1918*, pp. 476–88.

34. Vera Katsarkova, 'Ograbvaneto na Bulgariia ot germanskiia imperializm' [The Exploitation of Bulgaria by German Imperialism], *Trudove na V.I.I. Karl Marx*, vol. II (1969), pp. 164–223. For a brief survey, see Leo Pasvolsky, *Bulgaria's Economic Position after the War* (Washington, DC: Brookings Institution, 1930), pp. 53–60.

2 RECOVERY IN THE 1920s

Bulgaria's military defeat in 1918 left a legacy of recrimination which was to last for the entire inter-war period. Bitterness settled over Sofia like the city's winter fog. Political opinion became so divided that all sides were ready to resort to violence. The prospects for a multi-party system of Western parliamentary democracy were dimmer than ever, despite the hasty abdication and departure of Tsar Ferdinand. His son Boris assumed the throne in October 1918, at the young age of 24. Adding to the unsettled political situation were at least 220,000 refugees from the lost Macedonian lands and from previously Ottoman Thrace, now north-eastern Greece. They poured into the Pirin region on the Yugoslav border or the shanty town that quickly grew up in the western quarter of Sofia. Their divisions and irredentist frustrations were soon to make 'another Macedonian murder' a common newspaper headline in the capital city. By 1920, the burden of Allied war reparations hung over the entire country. The phrase 'bulgarska rabota', or a piece of Bulgarian work, began to spread in the Sofia cafés as a cynical synonym for something gone wrong because Bulgarians had laid their hands on it.

How, then, is one to explain the relatively strong domestic recovery and significant restructuring of foreign trade that characterised the Bulgarian economy during the 1920s? One reason surely was the country's compact character. It had no new territory or hostile ethnic group to assimilate from what had previously been Austria-Hungary. Such assimilation posed immense problems in greatly expanded Romania and newly constituted Yugoslavia and Czechoslovakia.

Bulgaria's political spectrum was, to be sure, neither compact nor converging. The divisions of the immediate post-war period only grew sharper over the next decade. The Bulgarian Agrarian National Union (BZNS) and Narrow Socialists, now the Bulgarian Communist Party (BKP), had been the parties consistently opposed to participation in the First World War. The two not surprisingly won half of the votes in the relatively free election for the Subranie in 1919, and fully 58 per cent in 1920. The Communists had increased their share to 20 per cent by 1920 at the expense of the fading Broad Socialists, now Social Democrats. The Agrarian shares of 31 per cent and then 38 per cent in the two elections had allowed them to form a government in October 1919 and to keep control in 1920. Yet the radical Agrarian regime of Aleksandur Stamboliiski had too many enemies and too few friends to survive beyond June 1923. Its violent overthrow began with Stamboliiski's brutal murder. An unholy alliance between army officers, leaders of the pre-war establishment and irredentists of the Internal Macedonian Revolutionary Organisation (IMRO) carried out the *coup d'état*. The failure of Stamboliiski and the Communist Party to cooperate in any way, now regretted by party historians, and the general hostility of Great Britain and France to the Stamboliiski regime over the reparations issues, left the Agrarians without any powerful allies of their own.[1]

The new regime broadened its base through a so-called Democratic Concord to include some five pre-war political parties. The Social Democrats decided to join this mixture of liberals and conservatives. Anti-monarchist Democrats stayed out. Agrarians and Communists were excluded. The BZNS soon split into *émigré* and resident factions, a division which was to plague the party for the next twenty years. The failed Communist uprising of September 1923 saw at least 5,000 party members killed and some of the party leadership forced into exile in the Soviet Union. Now illegal, the party was none the less able to establish the surrogate Bulgarian Workers Party in 1927.

Aleksandur Tsankov, a proto-fascist leader of both the June coup and the September defeat of the Communists, had meanwhile failed to retain his leadership of the Democratic Concord beyond January 1926. Opposition to his arbitrary methods hardened during the economic slump of 1925–6. The more moderate Andrei Liapchev now became Prime Minister and headed a series of coalition cabinets. Liapchev survived one resignation crisis in 1928 to rule until April 1931. He was himself born in Macedonia. His leniency, for whatever reason, allowed a resurgence of Macedonian violence and virtual IMRO autonomy in the Pirin region, including some collection of taxes.

Significant political differences separated the three leaders of these post-war regimes. Aleksandur Stamboliiski was an Agrarian populist. His opposition to the First World War had put him in prison for its duration. Aleksandur Tsankov had helped to organise the civilian Committee for Public Welfare during the war and wanted to remobilise Bulgarian society on the same military principles afterwards. Andrei Liapchev rejected the state's wartime ascendancy, on the basis of his own experience in the Ministry of War in 1918. He wished to rely instead on pre-war European principles of national liberalism. In different ways, each of these three men contributed to the growing Bulgarian tradition of political, although not religious or ethnic, intolerance.

Their economic policies had more in common. All three regimes were obviously struggling with the economic consequences of a lost war. All were without the consistent support of any of the victorious Western powers. In addition, the three leaders conceived of any solution to Bulgaria's problems primarily in economic terms. Unlike most of their counterparts elsewhere in south-eastern Europe, who were primarily lawyers, they had been trained in economics. Stamboliiski had studied agronomy in a German university and was a schoolteacher. He believed in private peasant property linked by the co-operative movement as the foundation for organising a better society. Occupational groups would be the basis for representational democracy.[2] To him, the huge peasant majority among the Bulgarian population justified his one-party regime. Tsankov was a fully trained economist who had risen to the rank of professor at Sofia University, a forum through which he continued to express his views after being forced from power. Liapchev had studied economics and finance during his university days in Zurich, Berlin and Paris. He was also a founder of the co-operative movement in Bulgaria, Minister of Agriculture and Trade for 1908–10, and as Minister of Finance in 1915, the major figure in unifying the growing number of popular savings banks. The primacy of economic programmes for these three men combined with continuing external restraint to create a more consistent set of post-war policies than their political differences would suggest.

The Shift from Wheat to Tobacco Exports

After the disastrous drought of 1918, more difficulties awaited Bulgarian grain cultivation during the post-war decade. Bad weather in 1921–2, 1924 and again in 1929 meant more bad harvests. The loss of the southern Dobrudja (see Map 1), as we have seen, had reduced the marketed wheat surplus by 20–25 per cent of the pre-war export level. This reduction cut the domestic wheat surplus from northern Bulgaria in half. The new territory in the southwest ran a grain-deficit, which claimed 30 per cent of the smaller domestic surplus in the south. What wheat and other grain remained available for export faced other constraints. Western European demand for cereal grain failed to regain its per-capita 1909–13 level until after 1925, and then did not exceed it.[3] Higher European incomes were being spent, not unexpectedly, according to Engels' law, on a more varied diet or on goods other than food.

Bulgarian grain had also lost some important competitive advantages in the remaining market. With Great Britain and France cut off from continental supplies during the war years, American and Canadian grain exports had risen sharply to meet the demand. These new sources of supply proved difficult for Bulgarian grain to displace. Railway rolling stock within Bulgaria had been worn out from heavy wartime use and lines further north from Yugoslavia had been damaged. Shipping on the Black Sea, and especially on the Danube, did not resume at pre-war levels for several years. The best Bulgarian port facilities were at Varna, now underutilised because of the loss of Dobrudjan grain. In addition, the absence of a national network of grain elevators for storing smallholder sales made quality control at an American or Canadian standard difficult.[4]

These post-war pressures accelerated the pre-war trend (noted in Chapter 1) away from exports of cereal grains, wheat in particular, and toward increased peasant consumption. Grain production per capita recovered to less than 90 per cent of the pre-war level by 1926–30, but cultivated grain area, yields and per-capita consumption all exceeded the 1909–12 norm on post-war territory. Table 2.1 makes the reason clear. The volume of grain exports per capita dropped to one-third of the pre-war level, the sharpest reduction for any of the Balkan states.

Tobacco exports more than filled the resulting gap. Table 2.2 reveals their striking rise from 1.3 per cent of total export value before the war to 38.5 per cent by 1926–30. This ascendancy had begun during the First World War (see Chapter 1). Cultivation on the newly acquired south-western lands was largely responsible (see Map 1). The pattern persisted after the war. As a result, the industrial crops, primarily tobacco, quadrupled their cultivated area to 3 per cent of the total. They increased from 3.3 per cent of total crop value in 1911 to 15 per cent by 1921–5. Such crops were still 11.9

Table 2.1: Grain[a] Production and Consumption, 1921–30 (1909–12 = 100)

	Area	Yield per hectare	Production per capita	Net exports per capita	Consumption per capita
1921–5	97	87	72	37	80
1926–30	111	104	89	33	103

[a]Wheat, corn, rye, barley, oats.
Source: J.R. Lampe and M.R. Jackson, *Balkan Economic History, 1550–1950* (Bloomington, Indiana: Indiana University Press, 1982), table 10.12, p. 365.

Table 2.2.: Composition of Exports, 1907–30 (in percentages)

	1907–11	1921–5	1926–30
Grain	55.7	23.4	14.5
Grain products	7.5	4.1	3.0
Other unprocessed crops	17.2	5.6	6.4
Tobacco	1.3	26.5	38.5
Rose essence	4.1	1.4	3.5
Eggs	7.6	8.1	12.4
Livestock	5.6	3.1	4.4
Hides	2.2	2.0	4.4
Other	8.8	25.8	12.9

Source: Lampe and Jackson, *Balkan Economic History, 1550–1950*, table 10.14, p. 368.

per cent for 1926–30, thanks to sugar beet and sunflower acreage, which together surpassed slumping tobacco cultivation. For the latter period, industrial crops were still 235 per cent of their real value for 1911, as against just 92 per cent for rebounding cereal grains. Total crop production in constant prices had climbed to 126 per cent of the 1911 level by 1926–30, a good recovery by European standards. Bulgaria was well ahead of neighbouring Yugoslavia at 90 per cent and Romania at 95 per cent.[5]

Bulgarian exports also grew ahead of the European average. The terms of trading primarily agricultural exports for primarily manufactured imports fell below the pre-war ratio. Yet the real per-capita value of exports was 47 per cent ahead of the 1906–10 level by 1926–30. Their average annual growth amounted to 2 per cent, compared with less than 1 per cent for Europe as a whole. The proportion of trade turnover (exports plus imports of goods) in Bulgarian national income had meanwhile, according to the rough estimates available for the latter aggregate, increased its one-

Table 2.3: Direction of Foreign Trade, 1906–30 (in percentages)

	North-west Europe[a]		Germany		Central and North-east Europe[b]		South-east Europe and Turkey	
	X	M	X	M	X	M	X	M
1906–10	37	29	12	17	7	30	35	19
1911–14	42	26	15	21	10	33	19	12
1915–18	3	9	42	32	37	33	9	19
1921–5	20	29	14	19	17	15	26	15
1926–30	18	26	25	22	25	20	14	11

X = exports; M = imports.
[a] France, Holland, Belgium, Great Britain, Switzerland and Italy.
[b] Austria, Hungary, Czechoslovakia, Poland, USSR or predecessors.
Source: Lampe and Jackson, *Balkan Economic History, 1550–1950*, table 10.13, p. 366.

quarter share of the pre-war period past the average fraction of one-third recorded by the developed European economies before the First World War.[6]

This promising picture still included the same sort of overdependence on a single export that had threatened pre-war prospects. True, processed tobacco accounted for less than 40 per cent of 1926–30 export value, in contrast to the 70 per cent of pre-war cereal grains. For the later period, however, grain exports had fallen under 15 per cent of the total. Other unprocessed crops, mainly turnips and kidney beans, had declined sharply to 6 per cent, according to Table 2.2, and livestock exports continued to be less. Only egg exports had continued their pre-war ascent, and were over 12 per cent of the 1926–30 total.

One consequence of the new overdependence on tobacco exports was an increasing reliance on trade with Central Europe in general, and Germany in particular. More groundwork, beyond the special ties of the First World War, was thereby laid for Bulgaria's close economic relationship with Nazi Germany during the 1930s. Table 2.3 calls our attention to the fact that by 1926–30, one-quarter of Bulgarian export value went to Germany and another quarter to the rest of Central and North-east Europe. There was no trade with the Soviet Union until 1934, and none of significance until 1940.

Once again, as with wheat exports, it was the Bulgarian peasantry which reacted more quickly than commercial interests or government policy to the dangerous new dependence on tobacco. A

Table 2.4: Sources of Population Change, 1906–40 (annual average per 1,000)

	Births	Deaths	Natural increase	Actual increase	Implied net migration[a]
1906–10	42.1	24.0	18.1	14.4	−3.7
1911–20	—	—	9.1	10.5	+1.4
1921–5	39.0	20.8	18.2	20.4	+2.2
1926–30	33.2	17.9	15.3	14.5	−0.8
1931–5	29.3	15.5	13.8	12.0	−1.8
1936–40	23.3	13.7	9.6	7.8	−1.8

[a] Calculated as the difference in actual and natural increase.

Source: Lampe and Jackson, *Balkan Economic History, 1550–1950*, tables 10.2 and 12.1, pp. 333, 437.

bad Bulgarian harvest in 1925 coincided with the full-scale return of Greek and Turkish tobacco to the European market-place. The subsequent revival of their exports reduced the European price by one-third for 1925–6. Bulgarian peasant growers did not wait for recovery. In 1926, they reduced the land under tobacco cultivation by one-half from the peak level of 1923. European prices had risen by 1927, thus permitting Bulgarian tobacco exports for 1926–30 to increase by 5 per cent over 1921–5, and to record the substantial increase in their share of total export value seen in Table 2.2. At the same time, the permanently reduced hectarage caused tobacco output, as reflected in the constant price index of industrial crop production, to decline by one-fifth over the same period.[7]

Peasants also reacted to the uncertainties of the European market by reducing their rate of population increase, although their proportion of total Bulgarian population remained overwhelmingly large during the 1920s. Those dependent on agriculture still stood at 74 per cent of the 6 million total in 1930, as against 75 per cent of 4.4 million in 1910. By 1930, only 12 per cent resided in towns over 20,000. Rural reluctance, rather than rapid urbanisation, was thus responsible for cutting what had been the highest pre-war birth rate in the Balkans from 42 per 1,000 to 33 per 1,000 between 1906–10 and 1926–30, and to 23 per 1,000 by 1936–40. Even a falling death rate could not prevent the decline in natural increase noted in Table 2.4.

It would be wrong to attribute all this decline to reaction against market reverses, real or anticipated. War losses and the fear of overdivided holdings also played their part. All sons inherited equal shares of the father's land. The density of population in 1910 on what would be Bulgaria's post-war territory was only 42.3 people per kilometre, the lowest Balkan level. Its increase to 55.1 people per kilometre was the region's sharpest increment, aided by relative absence of civilian deaths from war and disease during 1912–18 and the subsequent influx of refugees.[8]

The reduced rate of rural births was at least in part a sign that the peasant smallholder was losing confidence in his ability to support a household of pre-war size during the 1920s. That this could occur in a period of rising agricultural exports casts doubt on the capacity of Bulgarian economic growth in the 1920s to sustain itself even without the arrival of the Great Depression.

Smallholders and Stamboliiski's Legacy

Throughout the decade following the First World War, the three Bulgarian governments perceived agriculture to be in a state of almost permanent crisis. Finding land for the refugees from Macedonia and Thrace was a continuing problem. Outside aid to support their rural resettlement did not come until a 1926 loan from the League of Nations. Its dispersion was spaced out over a period of five years. Agricultural policy meanwhile revolved around how to improve the viability and efficiency of smallholdings. The journal of the Bulgarian Economic Society repeatedly called the modernisation of smallholdings the biggest task for economic policy in general.

Inter-war data collected by the Rome Agricultural Institute and by a Bulgarian economist suggest that the smallholdings' capacity for efficient performance and a marketed surplus was not invariably inferior to that of large-scale holdings, as Marxist scholars, among many others, have tended to assume. The Bulgarian record suggests that smallholdings near urban areas showed several modernising tendencies. They cultivated more labour- and even capital-intensive crops, and marketed a larger share of them than all but the largest properties far from town.[9]

Whatever their efficiency, smallholdings became even more predominant during the 1920s than they had been before the war. Stamboliiski's land reform was largely responsible. Political motives undoubtedly played their part. The notion of 'labour property', restricted to the size of holding which one peasant household could work, was after all central to Stamboliiski's Agrarian ideology.

The reform promulgated in the course of 1920 was, however, the result of more careful preparation and economic calculation than Western observers have typically assumed.[10] Co-operatives discussed the shape of the measure throughout the second half of 1919. Its provisions did decree the confiscation of absentee ownership over 4 hectares and provided for distribution of state land to households with less than 1 hectare. Yet its maximum holdings of 30 hectares provided exemptions for owners who could promise conversion to fruit or vegetable cultivation or to some form of manufacture within three years. Additional maxima of 20–50 hectares for pasture and forest were also specified. The uncertainty of title that would plague other Balkan land redistributions throughout the 1920s lasted only until 1922 in Bulgaria. From then, the Ministry of Agriculture paid former owners their promised partial compensation. Fallow land accordingly returned by 1924 to its pre-war proportion of cultivated area, about 17 per cent, compared with 23 per cent in 1921.

Only 330,000 hectares, or 4 per cent of arable land, was thereby redistributed, compared with 30 per cent in Romania. Nearly two-thirds of that 4 per cent was in fact state land. By 1926, holdings under 5 hectares covered 23.6 per cent of arable land, 5–10 hectares fully 34.5 per cent, and 10–30 hectares 36.6 per cent. The remaining 5.3 per cent over 30 hectares compared to 14.3 per cent in 1908.[11] Most of these previously large holdings were subdivided into strips for share-croppers. The land redistributed to peasant owners did not promote the consolidation of existing strips, which would have greatly facilitated more efficient cultivation, especially of grain.

The lack of facilities for grain storage also continued to hurt smallholders throughout the 1920s. Stamboliiski's grain consortium of 1920–1 had attempted to remedy this deficiency. Formed with funds from the state banks, the consortium was supposed to buy peasant grain at a fixed price and hold it off the market until a higher price could be obtained. Stamboliiski intended to use 90 per cent of the profits for constructing storage silos. The monopoly power given to the consortium's central purchasing bodies so provoked European grain traders that they persuaded the Allied Reparations Commission to force the Agrarian government to disband the system by 1921. Even before then, peasant sellers had won the right to 60 per cent of all profits. Allied permission allowed sizeable export duties to supplement state revenue instead, thus hampering the Bulgarian return to Western European markets still further.[12]

Despite the failure of land consolidation and the grain consortium, Stamboliiski's Agrarian regime left a series of useful legacies to Bulgarian agriculture. Practical education received special attention. Faculties of agronomy and veterinary medicine were established at Sofia University. Agronomists gave priority to new fodder seeds and stall-bred cattle. Secondary schools now taught more natural science and accounting. Students were also obligated to contribute two 'labour weeks' every summer, typically to help with the harvest, a practice which the present Communist government perpetuates. The programme for compulsory labour service, as implemented and revised in 1920–1, also stressed practical education. All males reaching 20 years old were drafted as *trudovaki* into uniformed labour battalions for one year's service. Some 30,000 were assembled the first year. Stamboliiski saw the service as something more than a device by which to instil social discipline and to circumvent treaty limits on a standing army. It became, as he intended, an institution for agricultural and industrial training. Until its loss of manpower in 1934 to an enlarged army, the service also constructed more bridges, roads, railway lines and economic enterprises than its support would cost the state budget.[13]

The Bulgarian Agricultural Bank (Bulgarska Zemedelska Banka) and the rural credit network of the Central Co-operative Bank emerged during the Stamboliiski regime as the largest sources of short-term credit in the country. Nearly 400 new credit co-operatives were formed, bringing the total to 1,200, and membership to over one million. Loans, which had been less than the Narodna Banka's in 1911, were now three times that total. Together, the Zemedelska Banka and the co-operative network had assembled the largest assets of any set of agricultural credit institutions in south-eastern Europe. The industrial crops of tobacco, sunflowers and sugar beets received short-term credit to improve plant selection and to introduce artificial fertiliser. Loans for these two purposes plus irrigation accounted for 40 per cent of the Zemedelska's credits to co-operatives. Several electric power projects received long-term credit.

Yet the system had its limits. Mortgage loans from the Agricultural Bank were inadequate; they declined to just 6 per cent of the bank's assets. The co-operatives themselves, moreover, gave three-quarters of their credits to payments of debt, purchases of land, and construction of new buildings, rather than improving agricultural efficiency directly.[14] Most of the co-operatives' activity was uncoordinated.

These limitations continued after the fall of Stamboliiski. Despite the formation of a central union of 50 tobacco co-operatives in 1925, members continued to undercut each other in seeking foreign buyers. Too much credit continued to go to rural housing. Funds were sometimes lost by corruption or, as in the case of a large electric power station near Plovdiv, by inefficiency. The divisions within the defeated Agrarian Party also began to show up among co-operatives. Credit organisations aside, agricultural insurance societies continued to grow, approaching 700 by 1928, but marketing and especially producers' co-operatives lagged behind. Only a few of the 50 producers' co-operatives had banded together to form 'rings' for buying tractors and other modern equipment.[15]

Perhaps more important, though, the whole co-operative apparatus survived and grew in assets, if not efficiency, under the regimes of the Agrarians' bitter political adversaries. Their publications pretended that the Stamboliiski era had never happened, but their policies declined to dismantle its institutions. The Tsankov regime did not seriously damage the co-operatives. In June 1924, it increased the general maximum for agricultural holdings by only 5 hectares per person in households over four, exempted vineyards and gardens from expropriation, and allowed some exceptional farms up to 150 hectares if they were 'modern and rational'.[16] But the redistribution of land continued under these reduced provisions and exceeded the total under the Agrarian regime. Exemptions from compulsory labour service could now be purchased, and the term of service cut from one year to eight months. Its annual levy dropped from 30,000 to under 20,000.

The Liapchev regime of 1926–31 had at its head an early promoter of the co-operative movement. Andrei Liapchev had opposed Stamboliiski's grain consortium of 1921–2 for political reasons, but had favoured encouragement for industrial or new crops and more intensive cultivation since before the war. The fall in tobacco prices none the less revived interest in grain cultivation. His regime took special interest in promoting the import of iron ploughs and steam-powered machinery through the credit facilities of the Agricultural Bank. As a result, the number of iron ploughs in use increased by 40 per cent for 1925–9, and their share of all ploughs approached the same percentage. Modern agricultural equipment more than doubled. Under Liapchév as well as under Tsankov, the Ministry of Agriculture continued to receive a larger share of state budget expenses (still only 3 per cent) than their counterparts elsewhere in south-eastern Europe. The several state experimental stations promoted better fodder crops and introduced selected seed to one-third of the area sown in wheat. Similar efforts were also under way for cotton cultivation, which was previously not of sufficient quality to permit mechanical spinning.[17] New construction also added 300 km of railway lines to a network now totalling 2,440 km.

Much remained to be done by the end of the decade. All three regimes, however, had given more aid to agriculture, directly through state policies and indirectly through the Agricultural Bank and the co-operative network, than had any of their Balkan neighbours. No comparable bank was even established elsewhere in south-eastern Europe until 1928.

Burden of Reparations

Bulgarian agricultural policy had to operate within a national economy whose financial resources as a whole were severely limited. When deceptively large nominal sums are adjusted for an 11.75-fold depreciation in the international value of the leva, real state budget revenues and expenditures per capita fell by 22 per cent between 1911 and 1920. After the leva's further depreciation to 37-fold, and slight improvement to a stable 26.8 times the 1911 parity with the French gold franc, we find real budget totals just reaching the pre-war level again during 1926–30. In the process, as Table 2.5 makes plain, note issues by the Bulgarian National Bank had been cut back by more than one-half from their 1920 level. The real per-capita average now barely matched the 1911 figure.

Behind these limitations lay two sets of pressures from the only source that might have offered the Bulgarian economy financial relief — the Western European community of bank and monetary officials. First, they pressured the Bulgarian government to pay its pre-war debts and the war-related reparations imposed by the Paris peace conference. Then they encouraged the Bulgarians to follow a

Table 2.5: Main Financial Indicators, 1911–30 (in constant 1911 leva per capita)

	State budget revenues	Exports	BNB note emission	State debt	Debt excluding reparations
1911	46	33	25	137	137
1920	35	39a	59	750	232
1926–30 (average)	47	42	25	177b	147b

a 1921–5.
b 1930 only.

Source: Lampe and Jackson, *Balkan Economic History, 1550–1950*, tables 11.2 and 11.3, pp. 384–7.

restrictive monetary policy, which would assure the exchange rate of the currency in which repayment was made.

Bulgaria was the only Balkan state on the losing side in the First World War. Afterwards, its neighbours joined the French-led coalition to make the losers pay. The threat of these new and old debts depreciated the exchange value of the leva sevenfold between 1919 and 1923, far more than the twofold domestic inflation would have dictated. The Stamboliiski regime chose to delay debt payments, thereby incurring the unremitting hostility of the Western European financial community.[18] French bondholders and officials were particularly sensitive to the possibility of another Russian-style repudiation. Lenin's Bolshevik government had already refused to pay any of the massive sum due on pre-war and wartime loans.

Reparations and related debts imposed by the Treaty of Neuilly in 1919 proved a more difficult problem. The total was far larger than pre-war obligations. With these impositions, according to Table 2.5, the state's foreign debt was over five times the 1911 level in constant foreign-exchange values. Without them, the increment was just 69 per cent. The initial levy consisted of costs for Allied military occupation and some service on the Ottoman public debt as well as reparations. The total came to 2,250 million gold francs, plus 5 per cent interest on the unpaid principal. It was due in semi-annual payments over 37-years beginning 1 January 1921. This fantastic sum amounted to perhaps twice the country's national income in 1911. The two payments a year would have totalled half of the state budget revenues. Its size, however, was consistent with the impositions levied against Germany, Austria and Hungary.[19]

So another French sensitivity, this time to make the Germans pay under the related Treaty of Versailles, put Bulgaria in a poor position to protest.

The Stamboliiski regime protested all the same. Payments in kind, mainly coal and livestock, were made to Yugoslavia and Greece. By 1922, their estimated value exceeded 100 million gold francs. Stamboliiski refused, however, to make the first cash payment to the Interallied Reparations Commission in Paris. Their representatives came to Sofia by February 1921 in order to extract payment. The government was able to put them off for over a year by agreeing to the aforementioned resumption of payment on the pre-war debt. The Commission continued to insist on tying all Bulgarian customs revenues to repayment of the full amount. The impasse lasted until 1923. Once the neighbouring Balkan states had failed to follow a French and Italian call for armed intervention, the Commission finally agreed to a reduction in what was owed. Only 550 million gold francs were now due over 60 years. Customs revenues still guaranteed payment; thus there was continuing pressure on the new Tsankov regime to maintain the damaging export tariffs.[20]

By the end of 1929, Bulgaria had paid a total of 41 million gold francs in reparations and related claims, in addition to a sum in excess of 100 million gold francs paid earlier in kind. The cost of these payments for the Bulgarian government came exactly to the sum of budgetary deficits for the period 1924–9. Annual obligations amounted to about 5.5 per cent of average revenues.

The price to the Bulgarian economy was larger than this relatively modest percentage might suggest. First, the threat of much larger payment between 1920 and 1923 had triggered a depreciation in the leva's exchange value that brought it down from 11.75 to 37 times the amount required to buy a Swiss franc in 1911. Second came the pressure to continue export tariffs. Most damaging, however, was the Commission's subsequent deflationary pressure on the state budget to minimise expenses so as to assure debt repayment, and on the Bulgarian National Bank to make the leva convertible into gold or Western currency at a stable rate of exchange. This gold exchange standard was achieved, informally in 1926 and formally by 1928, at the cost of the most drastic deflation of a money supply anywhere in southeastern Europe.[21] In order to understand the consequences of this deflation, we must first see how the Debt Commission and the League of Nations used new

Western loans to ensure it.

New State Loans, the National Bank and the Gold Standard

Returning to the pre-war gold standard was a goal pursued not just by Liapchev personally, but by the European monetary community as a whole. Most central bankers and treasury officials wanted it for their own countries too. They had succeeded, by the middle of the 1920s, at least in establishing a gold exchange standard which included the major gold backed denominations as well as gold for reserves. Several shortcomings of this ultimately unsuccessful attempt to revive the pre-war monetary system are well known: the shortage of funds for foreign investment and of gold itself, the weakness of the London capital market and the inexperience of New York's Wall Street. The most pervasive problem was the general desire of European governments, France's excepted, to stabilise their national currencies at a pre-war, and therefore overvalued, rate of exchange.[22] For Bulgaria, this problem took the form of a commitment to re-establishing pre-war parity with the French franc. In return for accepting an overvalued leva, the Bulgarians received only two relatively small Western loans whose funds were earmarked for purposes other than relieving the credit shortage. Liapchev hoped in vain that adherence to the standard would give the Bulgarian government access to international loans and 'room to live'.

Whether because of the general aversion of the European capital market to state loans or because of specific identification with the losing side in the war, the Bulgarian government was not able to present an acceptable request for a Western loan until 1926. In June of that year, the League of Nations' Financial Committee accepted the Bulgarian request for a loan to aid refugee settlement. The Committee tied its ascent, however, to strict conditions which the Liapchev regime reluctantly accepted. Most important was the Bulgarian side's agreement to bring the statutes of the National Bank 'into conformity with the best principles of central banking' as soon as possible. This meant the elimination of the bank's commercial lending, state loans included, and the maintenance of high reserves to protect the convertibility of the leva at 3.7 American cents, over one-third more than the low of 2.7 cents recorded in 1923. The loan of December 1926, nominally £2.4 million and US

$4.5 million, or £3.3 million total, and £2.9 million or 2,246 million leva realised, offered the National Bank no assistance. Some 1,600 million leva went to land and housing for refugee settlement in rural areas. The rest repaid Bulgarian treasury bills discounted by the Paris Bas Bank during the Balkan wars of 1912–13.

The League authorised a larger loan of £5.5 million in 1928. The £5 million realised amounted to 3,370 million leva. Some 30 per cent went toward paying off the state's past borrowing from the National Bank.[23] Rights to customs revenues that superseded the claim of reparations none the less secured the new loans, the conclusion of which also required the Bulgarian government to settle another pre-war debt, this time with the Disconto Gesellschaft Bank of Berlin for its 1914 loan. Germany's admission to the League of Nations now allowed it to press this claim for 85 million leva.

The repayment burden of these new loans and obligations was not high. They boosted the share of debt service in the state budget from 21 per cent to 24 per cent between 1926 and 1929, and the debt-service ratio to export earnings from 5.4 per cent to 7.7 per cent. The latter percentages were the lowest of any state in southeastern Europe. The greater disadvantage derived from the further limitations placed on the activity and currency emissions of the National Bank. Its discount rate therefore continued to be the highest among all the European central banks, except that of Greece. Its ratio of reserve assets to bank notes averaged 85 per cent for 1928–9, the highest in south-eastern Europe.[24] For 1926–30, according to Table 2.5 above, Bulgarian currency emissions in real per-capita terms had fallen back to the 1911 level.

Old Fiscal Problems

Before seeing how European banks made up for the deficiencies of the Western capital market, the reader should understand why the Bulgarian government could not do so itself. Simply saying that European financial orthodoxy stood in the way of the three postwar regimes does not provide a sufficient explanation for the budget deficits which plagued them all. The Stamboliiski regime ran huge deficits which were half of total revenues in 1919–20 and one-third in 1921–2. The Liapchev regime was responsible for deficits throughout the period 1926–30 that averaged one-fifth of revenues. Had unlimited credit from the National Bank, on the pre-1914 Bulgarian pattern, allowed these deficits to be larger still, the sort of catastrophic internal inflation which reparations triggered in Germany by 1923–4 might have overtaken Bulgaria as well. At their existing level, moreover, the Bulgarian deficits were in some measure the result of wasteful expenses and inefficient tax collection.[25]

Already large before the First World War, the state bureaucracy had almost doubled during the 1920s to reach 87,000 employees by 1930. The military, still limited by the terms of the peace treaty, were only 7,000 of this total. Together, their salaries accounted for 39 per cent of 1930 budget expenditures. Their numbers meant that 17 of every 1,000 Bulgarians were state employees, a figure four or five times the Western European average and surpassed in southeastern Europe only by the bloated Romanian bureaucracy. The honesty and efficiency of the Bulgarian bureaucracy were also suspect, though again, not as much as the Romanian. The crime rate among state employees rose by over 400 per cent from 1910 to 1926, compared with an increase of 64 per cent for the total population. By 1928, almost one-quarter of all Bulgarian crimes were committed by state or local officials. These abuses extended, for instance, to funds from the refugee loan. Houses were cheaply built and roads or bridges overpriced. The lower level of officials remained inadequately trained and paid; those at high levels were well paid, but often worked just a few hours a day.[26]

Revenue collection had increasingly come to rely, as we have seen, on indirect taxes. Customs and a variety of excise taxes were responsible for over half of budget revenues by the early 1920s. When paid, the major excises on tobacco, alcohol and sugar were regressive. In rural areas, especially in the Pirin region controlled by IMRO, they were either not paid, or privately collected with the government only given a fraction of its due. No proper procedure existed even to audit the amount of this lost revenue. The Stamboliiski regime had tried to introduce a new set of direct taxes that would collect 8–50 per cent of trade or industrial profits. Tsankov cut those percentages back. A bigger problem with these levies and the 4–10 per cent personal income tax was actually collecting them. Bribes to tax officials permitted widespread evasion. If the 2 billion leva cited in one estimate as outstanding direct taxes for 1928 had been collected, they would have more than covered all the budget deficits for 1926–30.[27]

At least in Balkan perspective, the revenues actually collected and spent by the Bulgarian government were better distributed. Economic and educational expenditures accounted for 32 per cent of the 1929–30 budget, a figure approached only by Yugoslavia and three times the Greek proportion. Among the economic expenditures, agriculture received the largest share of any Balkan budget, over 3 per cent of total spending. Some 14 per cent went for education and contributed to keeping the Bulgarian system of secondary education the best attended in south-eastern Europe.

Western European Banks in the Financial Breach

Although steering generally clear of state loans through bond issues, Western European banks none the less stepped in during the 1920s to fill the public and private breach in Bulgaria's financial resources. These banks provided short-term credit to both government and commercial borrowers, the latter largely industrial, as we shall see. The total covered 81 per cent of the sizeable deficit on current account plus reparations for 1924–6.[28] Repayment of these short-term obligations then absorbed virtually all of the additions to capital account from the two state loans of 1926, effectively 1927, and 1928. More short-term credit from these same banks then flooded in to help cover the record current-account deficit of 1929.

Soon after the First World War a number of these Western European banks had opened affiliates in Sofia. The now mainly French Balkanska and Generalna, plus the German Kreditna Banka remained from the five cautious foreign banks that had appeared in Sofia during the last pre-war decade. The newcomers were primarily from France, Italy, Belgium and Switzerland. They typically sought and attracted sizeable Bulgarian capital. Together, these 13 banks accounted for about 40 per cent of private bank capital in Bulgaria by 1929. The foreign share of their capital was 38 per cent. Thus we may speak of Western European shareholders actually owning about 15 per cent of private Bulgarian bank capital, about the same as their share of industrial capital.

This modest percentage understates their actual influence in Bulgarian banking. Direction, if not management, was almost always in these Western European hands. Their assets amounted to over 60 per cent of the private Bulgarian total. They were responsible for bringing the real per-capita assets of private commercial

Table 2.6: Loan Value by Bank Type, 1911 and 1928 (in millions of constant 1911 leva)

	1911	1928
Private commercial banks	125	239
Central State Bank (BNB)	126	48
Agricultural Bank (BZB)	113	119
Central Co-operative Bank	2	14
Popular banks		53
Agricultural credit co-operatives		18
Total	366	490

Source: Lampe and Jackson, *Balkan Economic History, 1550–1950*, table 11.5, p. 397.

banks back to the 1912 level by 1928. The public network of agricultural and co-operative banks could recapture only 49 per cent. The respective real levels of loans from private as against agricultural banks, compared to their 1912 levels, also favoured private institutions, 89 per cent as against 58 per cent.[29] Table 2.6 shows their absolute advantage. With the relative decline of the central bank, the overall balance of loanable funds had shifted towards private, primarily European banks by 1928.

The activities of these European affiliates had also broadened since the pre-war period. They continued to credit the principal European export traders and insurance firms, most prominently the French Dreyfus for grain, the English Robinson-Anderson for eggs, and the Italian Guardini for poultry. The fact that three-quarters of the foreign-owned trading firms were French, Belgian or Italian did not prevent the continuing growth of exports and imports to Central Europe, as recorded in Table 2.3. A multilateral system of foreign trade had thereby emerged, and from 1926 a system of multilateral payment. These multilateral activities, freer from political manipulation by the respective European foreign offices than before the war, extended to industry as well.

Seven European-backed banks, plus three entirely Bulgarian ones each had networks of industrial and trading enterprises which they supported through purchase of some stock, typically less than half, and through short-term credit on current account. Short-term support predominated by a ratio of perhaps two to one. Direct investment or sole ownership were rare. Complaints about the high cost of current-account loans and about the shortage of funds for long-term industrial investment appeared regularly in the journal of the Bulgarian Economic Society. The combined bank networks none the less included 51 manufacturing or mining firms and 55 in trade or insurance.[30]

Each European bank supported a sugar refinery and other food-processing enterprises. Coal and copper mining were left to direct European investment. Textiles and metallurgy were almost entirely neglected. The half-dozen firms of the small German Kreditna Banka still included the large Granitoid cement works of Sofia that it had helped found before the First World War. The Franco-Bulgarska Banka inherited the United Tobacco Factory, a tobacco-processing and sales cartel, from the Generalna. The three Bulgarian banks with such networks had ties to ten industrial firms whose capital was less than one-fifth of the foreign total. European banks had now become the principal source of credit for some branches of large-scale Bulgarian industry, a role which pre-war counterparts had failed to play.

Industrial Growth without Development

Mechanised manufacturing apparently grew at an impressive rate during the 1920s. Considered only in connection with the influx of European bank credit, this growth suggests the start of sustained development, including structural shifts of labour and capital. Only the depression of the 1930s was to call a halt to their development, according to this view. Available evidence, however, argues otherwise.

The growth rate of industrial production was at first glance spectacular. Real net output for all manufacturing enterprises exceeding 10 employees and 10 horsepower rose by 13.3 per cent a year for 1921–31. Its price-adjusted value doubled to reach 301 per cent of the 1911 figure by 1931. The great state-encouraged majority averaged a 6.8 per cent increase a year for 1909–29, and an annual 15.6 per cent increase for the post-war decade, according to Tables 2.7 and 3.4. These rates were the highest in south-eastern Europe. So were the rates of growth for industrial labour and horsepower, averaging 11 per cent for 1921–9, compared with 2–4 per cent for Yugoslavia and Romania. The real value of capital investment recaptured its 1912 level by 1924 and rose 22 per cent a year to more than double by 1929. The issue of joint-stock capital for industrial firms had become common practice since the last

Table 2.7: Structure and Growth of Industry, 1911–31

		Net output (%)[a]		Average annual growth (%) 1909–29[a]
	1911	1920	1930	
Metals and machinery	5.0	3.6	8.2	7.4
Chemicals	4.5	4.0	10.5	13.5
Construction materials	7.3	8.0	9.1	10.1
Wood processing	3.9	1.6	2.5	1.5
Textiles	29.4	16.6	22.3	6.5
Leather	2.6	3.0	2.3	1.4
Foodstuffs	46.2	62.3	43.8	6.0[b]
(Flour)	(14.8)	(20.4)	(9.7)	(1.4)
Total				6.8

[a] Private encouraged enterprises.
[b] Approximate.

Sources: Lampe and Jackson, *Balkan Economic History, 1550–1950*, table 11.9, pp. 408–9; M.R. Jackson and J.R. Lampe, 'The Evidence of Industrial Growth in Southeastern Europe before the Second World War', *East European Quarterly*, vol. XVI, no. 4 (1983), p. 396.

years of the First World War, at least in Sofia, where over 60 per cent was concentrated. The number of incorporated manufacturers had jumped from 37 in 1909 to 128 in 1921 and 263 by 1930.[31] In virtually every branch of industry except metallurgy and machinery, at least a few large companies produced their wares according to European standards of best practice for both technology and organisation. Yet these few firms failed to extend their best capitalist practice to the rest of Bulgarian industry. Nor did the structure of industrial production change significantly. As noted in Table 2.7, metals and machinery were still only 8 per cent of output in 1930, textiles and foodstuffs 66 per cent.

One reason was surely the limited attractiveness of industry, especially heavy industry, to private investors, both foreign and domestic. Profits from joint-stock manufacturing continued their pre-war pattern by lagging several points behind the 10 per cent averages for banking and commerce. Some 40 cartel arrangements appeared during the 1920s, in contrast to a half-dozen before the war. Most survived only a few years, but even the attempt to assemble them speaks poorly for the prospects of continued growth. Joining the tobacco cartel in sustained effectiveness was a sugar cartel. It included the above-mentioned refineries supported by the European bank affiliates. Together, they virtually eliminated imports, but operated at only two-thirds of capacity. They priced sugar so high that the processing of jam and other potential food exports was discouraged. All branches of manufacturing, unlike mining, still found long-term European investment hard to attract. Foreign capital accounted for about half of shareholdings in the small private sector for mining. It accounted for perhaps 15 per cent in manufacturing, owning 45 per cent of the stock in 15 per cent of the firms.[32]

The new supply of short-term, current-account credit from the European bank affiliates helped to relieve the long-term shortage, although not to the extent of financing expensive new European machinery. In any case, the approximate halving of the National Bank's commercial credit and also of the share given to industry in the later 1920s kept loans from the European affiliates expensive, with an interest rate of 10–15 per cent.

The small size of the domestic market was another continuing problem. True, Sofia had more than doubled its pre-war population, rising to 230,000 by 1926. The new residents were now predominantly refugees or, from 1925, peasants from the tobacco-raising south-west. The lateral movement from other Bulgarian towns to the capital city was less important than before the war. Yet the urban share of Bulgaria's total population hardly rose by a significant amount between 1910 and 1930, from 19 per cent to 21 per cent. The share of the population not dependent on agriculture, and presumably more inclined to buy manufactured goods, rose even less, from 25 per cent to 26 per cent.[33]

A final obstacle to sustained development, whether under private or state auspices, was the small initial size of the modern industrial sector and of the average factory. Employment in this sector more than doubled to reach 55,000 by 1921 and perhaps 75,000 by 1929. Rural and urban artisans continued to outnumber this factory employment, so that total 'industrial' employment rose less, from 180,000 to 270,000 from 1910 to 1929, and increased its share of the entire labour force minimally from 12.5 per cent to 13 per cent. Equally important, the size of even the encouraged enterprises stayed the smallest in south-eastern Europe. The average number of workers per factory rose slightly between 1921 and 1929, from 36 to 41, and exceeded 100 only for textiles. Horsepower per worker declined slightly to 2.56.[34] Small enterprises, just meeting the minimum requirements for tax and tariff exemption, continued to appear. Partnerships and even individual proprietorships that met the minimum requirements had grown from 214 in 1909 to 326 in 1921 and then almost tripled to 899 by 1929. As a result, the share of joint-stock enterprises actually fell during the decade, from 23 per cent to 21 per cent.[35]

In order to attract much capital and labour from the rest of the Bulgarian economy, modern industry would have had to have grown at a fantastic pace, under any circumstances. The best Bulgarian estimate of national income before and after the First World War, by Chakalov, reckons that real per-capita growth was literally nil between 1911 and 1926, with perhaps some slight advance from 1926 to 1927.[36] Modern industry admittedly showed the most rapid growth of any sector, but its increase from 1911 to 1926 could lift its share of national income only from 2.7 per cent to 5.1 per cent. This just surpassed rural and urban artisan production, which fell from 6.8 per cent to 5 per cent. Rural household manufacture, more difficult to estimate than rural artisanware, actually rose, from 7.7 per cent to 9.3 per cent. The peasant majority preferred, in other words, to manufacture its own margin above subsistence.

This diversion, noted both before and after the First World War, doubled the dilemma so familiar in other developing economies: how to assemble a large marketed surplus of primarily agricultural goods, and then invest the proceeds in modern manufacture so that the latter becomes the predominant sector, when it is so small a share of national income in the first place. By the late 1920s, the Soviet Union was offering one alternative: forced collection of the agricultural surplus and massive state investment in industry. Denmark was offering another: the increased processing of agricultural goods for export through the co-operative movement. At the time, Bulgarian governments rejected one approach and failed to adopt the other successfully.

Protectionism and Other Industrial Policies

The Bulgarian government's efforts to follow the Danish example were most noticeable during the Agrarian regime of Aleksandur Stamboliiski. From none existing before the First World War, the number of co-operatives qualifying as encouraged manufacturing enterprises jumped to 95 by 1921, nearly equalling the 128 private incorporated firms.[37] Stamboliiski's other economic policies confirm his pre-war disposition, noted in Chapter 1, to favour industrial development if tied to domestic agricultural inputs. In 1921, his Agrarian regime had also extended the pre-war legislation for industrial encouragement for another ten years. The incidence of his new direct taxes on enterprise income was purposely lower for industry than for commercial firms or property renters. New tariff legislation was expressly forbidden by the peace treaty. Stamboliiski did succeed in applying coefficients for currency depreciation, which kept pre-war rates partly in place. His regime also banned imports of a few specific luxuries and collected export tariffs on grain.

The Liapchev regime, buoyed by Bulgaria's better international standing, passed a new comprehensive tariff in 1926 and a new industrial law in 1928. On the one hand, the new import tariff laid down the sharpest set of increases recorded among 13 states of continental Europe during the 1920s.[38] Maximum levels tripled for foodstuffs between 1913 and 1927, doubled for semi-manufactures, and quadrupled for finished manufactures. As a result their *ad valorem* levels increased from 25 per cent to 60 per cent (75 per cent for finished manufactures). These were genuinely protectionist levels.

Their role in the rapid growh of Bulgarian industry during the late 1920s does not, however, appear to have been decisive. Coal imports had been eliminated by an increase in domestic production during the First World War and remained minimal afterwards. With the exception of sugar and cement, the highest levels of protection do not correlate with the highest rates of growth. Yet sugar production still did not exceed two-thirds of capacity. Cement would face little competition from imports anyway, given its weight/value ratio. The two fastest-growing branches of Bulgarian industry, chemicals and metallurgy, were relatively less protected. Overall, the ratio of import tariff revenues to import value rose from 13.3 per cent to 16.8 per cent for 1925–9, a small and unclear response to the 1926 tariff. More clearly, the high tariff levels made entry into most branches easy enough to help frustrate the great majority of the 40 manufacturing cartels established during the decade. Little wonder that they typically dissolved after one or two years, as already noted.

Further facilitating the easy entry which classically discourages concentration were the terms of Bulgaria's industrial encouragement laws. Liapchev's much-heralded new legislation of 1928 did not make any significant changes in the pre-war privileges which Stamboliiski had simply renewed, other than a 25 per cent cut in rates for rail freight.[39] The new law kept the minimum number of employees for qualifying enterprises at ten. The economies of scale and further mechanisation promised from enterprises of even close to that size were nil.

The principal economic effect of industrial encouragement probably lay elsewhere. Exemptions from the high new tariffs of 1926 and reduced rail rates stimulated a flood of new imports, not of finished manufactures, but rather of semi-finished inputs like cotton thread, construction materials and metal products. Overall, the imported fraction more than doubled during the 1920s to reach 40 per cent of total industrial inputs by 1930. Mainly because of metals, machinery and construction materials, this fraction exceeded the 29 per cent recorded by the Romanian economy, which had the largest industrial sector in south-eastern Europe.[40] This influx admittedly involved abuses. Bulgarian critics could, for instance, point to steel wire imported for nail manufacture or imported wool yarn that cost more than imported nails or cloth. There were even more reports than before the First World War of small encouraged enterprises using their duty-free privileges to import semi-finished goods, which they simply sold on the open market.[41] In addition, these imports put further pressure on the Bulgarian balance of payments.

Yet a strong case could be made for the propsition that this influx of inputs promoted the Bulgarian industrial development in the long run. The imported share of domestic industrial consumption as a whole dropped from 70 per cent in 1909 and 62 per cent in 1921 to 38.5 per cent by 1929. The reduction of finished imports was much greater than the rise in the semi-finished. The potential for abuse, had the first reduction been attempted without the second increase, may be seen in the actual case of cement, which was a major industrial input. The large Granitoid plant in Sofia took advantage of the absence of import competition, more because of the aforementioned weight/value than the tariff level, to pass on to customers the full amount of the tariff and thereby more than double the selling price. The encouragement laws and the availability through imports of most other industrial inputs otherwise combined to spare Bulgarian industry the burden of protectionism.

The rest of the Bulgarian economy bore the burden of higher

prices for finished manufactures. At the same time the rising production of finished manufactures and imports of semi-finished goods locked the economy into a new round of import substitution and export growth that would eventually encourage further industrialisation. One hopeful sign in the late 1920s was the success of the Agricultural Bank's pilot project for introducing improved cotton seeds.[42] Their cultivation was to produce cotton of sufficient quality to be used in textile manufacture. During the 1930s (see Chapter 3), this cultivation spread widely enough to reduce imports of raw cotton and cotton yarn.

By then, the world depression had unfortunately served to isolate the Bulgarian economy from multilateral trade and the short-term foreign credit that had helped to fuel the striking recovery of the 1920s. During the decade, however, peasant smallholders had already expressed their doubts about prospects for growth beyond recovery. They reduced their birth rates. They turned away from tobacco marketing, as they had from wheat. Multilateral trade did encourage the search for new exports. But the fragility of the European recovery during this post-war decade, and especially the failure of foreign trade to grow at a rate even approaching the pre-war pace, made the search a frustrating one for most small economies.

Notes

1. On the Agrarians' rise and fall, see John W. Bell, *Peasants in Power: Alexander Stamboliski and the Bulgarian Agrarian National Union, 1899–1923* (Princeton, New Jersey: Princeton University Press, 1977), pp. 122–246. On the Communist role and the subsequent failure of their September uprising, see Joseph Rothschild, *The Communist Party of Bulgaria: Origins and Development, 1883–1936* (New York: Columbia University Press, 1959), pp. 85–151.
2. Bell, *Peasants in Power*, pp. 55–84.
3. Louis G. Michael, *The Cereal Crop Situation in Bulgaria*, Technical Bulletin, no. 25 (Washington, DC: US Department of Agriculture, 1923), pp. 6–15; Wilfred Malenbaum, *The World Wheat Economy, 1885–1939* (Cambridge, Massachusetts: Harvard University Press, 1953), pp. 79–84; League of Nations, *Agricultural Production in Continental Europe during the 1914–1918 War and the Reconstruction Period* (Geneva, 1943), pp. 11–18.
4. Louis G. Michael, *Agricultural Survey of Europe: Danubian Basin, 2*, Technical Bulletin no. 126 (Washington, DC: US Department of Agriculture, 1929), pp. 93–8.
5. See in John R. Lampe and Marvin R. Jackson, *Balkan Economic History, 1550–1950: From Imperial Borderlands to Developing Nations* (Bloomington, Indiana: Indiana University Press, 1982), table 10.9, p. 359; Marvin R. Jackson, 'Agricultural Output in Southeastern Europe, 1910–1938', *ACES Bulletin*, vol.

XXIV, no. 4 (1982), table 3, p. 56. For a still useful survey, see Leo Pasvolsky, *Bulgaria's Economic Position after the War* (Washington, DC: Brookings Institution, 1930), pp. 193–213.
  6. See Lampe and Jackson, *Balkan Economic History*, tables 10.6 and 12.13, pp. 343, 480; Nikolai Todorov et al., *Stopanska istoriia na Bulgariia, 681–1981* [Economic History of Bulgaria, 681–1981] (Sofia, 1981), pp. 337–40.
  7. See Lampe and Jackson, *Balkan Economic History*, table 10.9, pp. 359, 371.
  8. Ibid., tables 10.1 and 10.3, pp. 332–4.
  9. See note 13 in Chapter 3 below.
  10. Bell, *Peasants in Power*, pp. 162–7, contradicts the long-established view of all Balkan land reforms, Bulgaria's included, as hasty political acts, typified by L.S. Stavrianos, *The Balkans since 1453* (New York: Holt, Rinehart and Winston, 1958), p. 594.
  11. Lampe and Jackson, *Balkan Economic History*, table 10.8, p. 357; Michael, *Cereal Crop Situation in Bulgaria*, pp. 7–10.
  12. Overseas Trade Department, Great Britain, *Economic Conditions in Bulgaria*, January 1922 (London: HM Printing Office), p. 95; Michael, *Danubian Basin*, pt 2, p. 95; League of Nations, *Chronology of Political and Economic Events in the Danubian Basin, 1918–1936, Bulgaria* (Paris, 1938), pp. 10–12.
  13. George C. Logio, in *Bulgaria Past and Present* (Manchester: Sherratt and Hughes, 1936), p. 165, notes that half of the value constructed by the service for 1922–31 was in bridges and roads, and 19 per cent in railways and port facilities, 21 per cent in industrial enterprises or sawmills. The standard source on the founding of the service is Max Lazard, *Compulsory Labor Service in Bulgaria* (Geneva, 1922).
  14. Lampe and Jackson, *Balkan Economic History*, tables 10.16 and 11.5, pp. 372, 396; J.S. Moloff, 'Bulgarian Agriculture', in O.S. Morgan (ed.), *Agricultural Systems of Middle Europe* (1933; reprinted, New York: AMS Press, 1969), pp. 69–79.
  15. Logio, *Bulgaria, Past and Present*, pp. 151–60; Lampe and Jackson, *Balkan Economic History*, pp. 369–71.
  16. G. Danailov, *Les effets de la guerre en Bulgarie* (Paris, 1932), pp. 566–72.
  17. Moloff, 'Bulgarian Agriculture', pp. 64–7; *Statisticheski godishnik na Bulgarskoto Tsarstvo* [Statistical Yearbook of the Bulgarian Kingdom], *1930* (Sofia, 1931), p. 346.
  18. William H. Wynne, *State Insolvency and Foreign Bondholders*, vol. II (New Haven, Connecticut: Yale University Press, 1951), pp. 544–9, provides the most concise and accurate summary of the debt and reparations negotiations.
  19. Ibid., pp. 549–51.
  20. Harold G. Moulton and Leo Pasvolsky, *War Debts and World Prosperity* (New York: Brookings Institution, 1932), pp. 289–90.
  21. Lampe and Jackson, *Balkan Economic History*, pp. 382–8, including table 11.2, and also table 12.13, p. 480; Pasvolsky, *Bulgaria's Economic Position*, pp. 115–47.
  22. See Derek H. Aldcroft, *From Versailles to Wall Street, 1919–1929* (Berkeley, California: University of California Press, 1977), pp. 125–86.
  23. Another 25 per cent went for railway construction, 10 per cent for relief from the 1928 earthquake, 22 per cent for budget arrears, and 13 per cent for the Agricultural and Co-operative Banks. Wynne, *State Insolvency and Foreign Bondholders*, pp. 552–6; Logio, *Bulgaria Past and Present*, pp. 81–107.
  24. See Lampe and Jackson, *Balkan Economic History*, tables 11.4 and 12.13, pp. 392, 480.
  25. The best treatment of this subject remains the pro-Zveno volume of Logio, *Bulgaria Past and Present*, pp. 28–49, 91. Bulgarian Marxist scholarship has paid it surprisingly little attention, concentrating instead on the growth of private industry

and European financial penetration.

26. See Lampe and Jackson, *Balkan Economic History*, tables 12.20 and 12.22, pp. 501–5.
27. Logio, *Bulgaria Past and Present*, p. 160; Todorov, *et al., Stopanska istoriia na Bulgariia, 681–1981*, pp. 331–2.
28. National Bank loans and long-term foreign investment in stock shares covered the rest. The only comprehensive calculation of the Bulgarian balance of payments during this period is by Marvin Jackson, in Lampe and Jackson, *Balkan Economic History*, table 12.25, pp. 512–16. Also see League of Nations, *Memorandum on International Trade and Balance of Payments, 1913–27* (Geneva, 1928) and *Balance of Payments, 1928–1939* (Geneva, 1929–39). A good Bulgarian survey is A.L. Georgiev, 'Otrazhenie na vunshnite finansovni dulgovi . . . 1918–1939' [The Impact of Foreign Financial Debts . . . 1918–1939], *Trudove na V.I.I. Karl Marx*, vol. II (1966), pp. 355–62.
29. See Lampe and Jackson, *Balkan Economic History*, tables 11.5 and 11.15, pp. 396, 430. Also see Asen Chakalov, 'Stokoviiat i bankoviiat kredit v Bulgariia' [Commodity and Bank Credit in Bulgaria], *Spisanie na bulgarskoto ikonomichesko druzhestvo*, vol. XXIX, no. 1 (1930), pp. 18–33.
30. The Balkanska Banka had the largest single network, some 24 firms by 1924, including five in manufacturing. For details on all banks, see Liuben Berov, 'Le capital financier occidental et les pays balkaniques dans les années vingt', *Etudes balkaniques*, vol. 2–3 (1965), pp. 139–69.
31. See Lampe and Jackson, *Balkan Economic History*, tables 11.8 and 11.14, pp. 404, 426; Todorov *et al., Stopanska istoriia na Bulgariia*, pp. 304, 381.
32. See Lampe and Jackson, *Balkan Economic History*, table 11.15, pp. 424–9. The most thorough Bulgarian treatment of cartels is Liuben Berov, 'Kum vuprosa na monopolisticheskite organizatsii v Bulgariia' [On the Question of Monopoly Organisations in Bulgaria], *Ikonomicheska Misul*, vol. 7 (1958), pp. 68–78; vol. 10 (1960), pp. 57–74.
33. See Lampe and Jackson, *Balkan Economic History*, tables 10.3 and 10.4, pp. 334–6. On the growth of inter-war Sofia, see John R. Lampe, 'Interwar Sofia vs. the Nazi-style Garden City: The Struggle over the Muesmann Plan', *Journal of Urban History*, vol. 11, no. 1 (November 1984), pp. 39–62.
34. Lampe and Jackson, *Balkan Economic History*, tables 10.4 and 11.12, pp. 336, 419; Marvin R. Jackson and John R. Lampe, 'The Evidence of Industrial Growth in Southeastern Europe before the Second World War', *East European Quarterly*, vol. XVI, no. 4 (1983), pp. 397–8.
35. Lampe and Jackson, *Balkan Economic History*, table 11.14, p. 426.
36. The uncertainty surrounding records of livestock and artisan production make precise estimates of national income difficult, but earlier Western estimates of an aggregate decline from 1911 to the late 1920s seem unwarranted. See Marvin R. Jackson, 'National Product and Income in Southeastern Europe before the Second World War', *ACES Bulletin*, vol. XXIV, no. 3 (1982), pp. 73–103. The Bulgarian estimates of Chakalov and others are summarised by the leading Bulgarian specialist on national income analysis, Petur Shapkarev, *Statistiko-ikonomicheski etiudi vurkhu narodnoto-stopanstvo na NR Bulgaria* [Statistical-Economic Studies on the National Economy of the P.R. Bulgaria] (Varna, 1982), pp. 121–6.
37. Lampe and Jackson, *Balkan Economic History*, pp. 426–7; Todorov *et al., Stopanska istoriia na Bulgariia*, p. 328.
38. Lampe and Jackson, *Balkan Economic History*, pp. 411–13; table 11.10 is extrapolated by Marvin Jackson from the one comprehensive conversion of specific European tariff rates into *ad valorem* percentages, H. Liepmann, *Tariff Levels and the Economic Unity of Europe* (New York: Macmillan, 1938), pp. 396–413.
39. League of Nations, *Chronology . . . 1918–1936, Bulgaria*, p. 29;

K. Bobchev, *Promishlena politika* [Industrial Policy] (Sofia, 1932), pp. 186–91.
40. Lampe and Jackson, *Balkan Economic History*, table 11.11, p. 416.
41. Logio, *Bulgaria Past and Present*, pp. 134–42.
42. Lampe and Jackson, *Balkan Economic History*, p. 411.

3 ISOLATION IN THE 1930s

The depression decade began for Bulgaria, like the 1920s, with unprecedented pressure from the international economy. Political turmoil followed in 1934. As in 1923, army officers joined civilians in overthrowing an elected government. As in the previous decade, three different regimes ruled Bulgaria. Again, they were bitter political opponents, but pursued economic policies that were surprisingly consistent. Most of the policies had in fact already been launched during the Liapchev regime of 1926–31.

The onset of the depression from 1929 forward doomed the chances of Liapchev's Democratic Concord in the relatively free elections of 1931. By then world agricultural prices had tumbled down by one-half. Bulgarian export earnings started a slide which was to reduce them to less than half of the 1929 level by 1933. In the absence of any more state loans, and in the face of a growing shortage of European bank credit, the Liapchev regime had little to show for its determined adherence to the gold exchange standard. The country's balance of payments was protected only because imports fell much more rapidly than exports. The results for industrial production and consumption were predictably disastrous. A large 'scissors', in Trotsky's phrase, opened up between the prices of manufactured and agricultural goods. These economic weaknesses fed public dismay over the long tenure of the pre-war politician, Andrei Liapchev, and over his failure to bring the Internal Macedonian Revolutionary Organisation (IMRO) under effective control. The way was thus cleared for a new coalition government.

The victory of the National Bloc over the ruling Democratic Concord in the election of June 1931 does not suggest a significant change of ideological disposition. Key participants in the winning coalition were the small Democratic Party and the so-called Vrabcha Agrarians, still unreconciled with the *émigré* Pladne Agrarians. Their coalition was none the less able to rally a majority of peasant voters plus those urban interests hurt by the reduction in imports and bank credit. The 66-year-old Democratic leader, Aleksandur Malinov, became Prime Minister, but was quickly replaced by his younger colleague, Nikolai Mushanov. Each was a pre-war politician and a lawyer by training, rather than an economist like Liapchev. Dimitur Gichev led the several Agrarians appointed to the new cabinet. The regime aspired to reflect peasant interests primarily, but could not overcome the continuing decline in agricultural exports and the opposition of *émigré* Agrarians.

The Bulgarian Workers' Party, surrogate for the illegal Communists, continued to be excluded from the government, despite increasing support from industrial workers. The party's victory in Sofia's municipal elections of 1932 was soon nullified. The moderate Mushanov regime also failed either to win over or to bring the IMRO under control. The Pirin region remained outside Sofia's effective control, for collection of taxes as well as for support in Balkan foreign policy. Disillusion with traditional political parties spread rapidly in this atmosphere.

The Zveno movement had already attracted a small, but dedicated following to its authoritarian ideology. Since its founding in 1927, Zveno had preached technical efficiency under a new, non-party regime. Its alliance with the same Military League of army officers who had helped to overthrow Aleksandur Stamboliiski proved decisive. Together they staged a bloodless coup on 19 May, 1934 and seized power from the Mushanov regime. The Zveno leader Kimon Georgiev became Prime Minister. Colonel Damian Velchev played the role of military strong-man behind the scenes. Under the somewhat contradictory slogans of scientific efficiency and 'national regeneration', they sought to pursue the domestic goal of economic modernisation. The existing political parties were called disruptive and were disregarded. In the absence of any effort to found a mass political movement, to set up its own militia, or to deify its leaders, the Zveno must be called authoritarian rather than fascist.[1] Its regime aspired to rule better through a more efficient state bureaucracy. The IMRO was now suppressed, and with surprising ease.

In foreign affairs, Zveno hoped for alliance with France, rather than Italy, and for reconciliation with Yugoslavia. The regime also established diplomatic relations with the Soviet Union; these were to remain unbroken throughout the Second World War. Yet no meaningful economic ties had been created with either the Soviet Union or France by the time that Tsar Boris and the military high command carried out their own coup, again bloodless, on 22 January, 1935. The Zveno regime had lasted less than a year. Its orientation in foreign policy was quickly abandoned, but some important economic legacies remained.

The energetic Boris was free of the effete pastimes and aristocratic pretensions that made his father such an unsympathetic figure to most Bulgarians. Although he was a moderniser, he did not restore the constitution of 1879, but ruled rather through cabinets of close advisers, led by a figurehead prime minister, the aged Kioseivanov. The Subranie selected its members from approved lists of 'non-party' candidates until an election was finally allowed in 1938, still without the use of party labels. Under these conditions, all party structures except that of the illegal Communists began to atrophy. Efforts to reorganise the state bureaucracy, particularly the economic ministries, continued. The ministries of agriculture, commerce and industry were combined into a single Ministry of National Economy.

Ties with France and the Soviet Union were not pursued, and Boris turned increasingly to Germany. This fatal alliance developed less rapidly before the Second World War than might be inferred from the tsar's background as liaison officer to the German army in 1915–18. Certainly there was scant desire to follow the Nazi example politically. If Boris and his associates had regarded the Zveno regime as too radical, how could they have accepted the sort of perpetual mass mobilisation and strident propaganda campaigns which characterised Hitler's Germany? As late as 1935, the tsar specifically rejected the 'absurd theories' and 'totalitarian methods' of the Nazi regime. Only Aleksandur Tsankov and his small movement, plus several even smaller groups under General Lukov and others, advocated a real fascist regime for Bulgaria. Boris subsequently appointed a few of their adherents to his cabinets, but with the intention (and effect) of weakening their movements rather than strengthening them.[2]

It was instead economic isolation that first pushed Bulgaria into the German orbit. Where else could the Sofia governments of the 1930s have turned? The depression had reduced Western European markets for agricultural exports still further. New loans were out of the question. The restrictions of the Reparations Commission of the League of Nations had ended in 1931 with the collapse of Bulgaria's capacity to pay. Adherence to the gold exchange standard became meaningless. The Western powers themselves abandoned it, rather than face further domestic deflation. Economic isolation from the USSR continued a pattern set under the tsarist regime. Russian exports to other countries during the 1920s had in any case continued to be largely grain, still a Bulgarian export and rarely an import. Only the German economy remained, and it was expanding, first with rearmament and then with the acquisition of Austria and the Czech lands.

In addition to widening relations with Germany, the other distinguishing economic feature of the 1930s was the growth of state initiative, now unchecked by the Subranie or by the League's Reparations Commission. The record of economic growth from 1934 forward gave the Bulgarian government no reason to change course. National income may have doubled. For all of southeastern Europe, it was a period for 'the genesis of étatism', in the apt phrase of the British economist Michael Kaser.[3] For Bulgaria, more precedents were set for state initiative than anywhere in the region.

State Export Controls and Co-operative Credit

The Agricultural Bank and the co-operative credit networks had continued to expand their activities throughout the 1920s. Although state export controls had essentially disappeared with the demise of Stamboliiski's grain consortium, the depression brought them back. One of the last acts of the Liapchev regime was to establish a central grain-purchasing agency in late December 1930. Known as *Hranoiznos*, or 'grain export', the agency began operations in February 1931. It did not lose its separate identity until 1948, after playing a part in the post-war consolidation of Communist power (see Chapter 5).

As originally constituted, Hranoiznos could authorise the Agricultural Bank, co-operatives or private traders to act as its agent in purchasing grain at prices above the fast-sinking international level. The agency could then sell it to domestic flour mills. Founding capital came from loans secured through the Agricultural and Central Co-operative Banks. Their representatives sat on the agency's administrative board, along with delegates from the state's Council of Ministers and the private sector's Chamber of Commerce and grain exchange. Under this mixed management, its powers expanded into those of a purchasing monopoly by October 1931. Purchases had become so large that in order to pay for them the agency needed to impose a surcharge on domestic flour sales. Only monopsony powers could ensure its effectiveness. Hranoiznos soon acquired this sales monopoly.

The agency had paid for only a fraction of each purchase in cash. Bonds good for payment of land taxes covered the rest. Yet agency prices were high enough to attract most of the bumper crops of 1930 and 1931. Almost all was then exported at a substantial loss, if land taxes foregone are included. The new Malinov-Mushanov regime had cut purchase prices by 15–20 per cent in August 1931, but world prices were falling even faster.[4] The regime thereupon decided to charge the urban consumer rather than the state budget for this programme of price supports. Major beneficiaries were peasant households with holdings over 10 hectares; they typically marketed half or more of their grain crops.[5]

From October 1931, the powers of Hranoiznos went beyond those of Stamboliiski's grain consortium. The agency's monopoly on domestic grain purchase allowed one of its salaried controllers to monitor the operation of each large commercial flour mill and of groups of smaller ones. Local milling for the grower's own consumption was monitored by the state-appointed district mayor. To aid supervision, the agency required mills to keep detailed records open to official inspection. Hranoiznos soon extended similar regulations to sugar-beet cultivation. From April 1932, the agency set out not only prices, but also designated the hectarage sown for each of the country's dozen refineries. The production was in turn divided into delivery quotas to meet domestic demand. This was really a state sugar cartel, replacing the private one which the Mushanov regime had officially dissolved the month before.

Hranoiznos also joined with the Agricultural Bank in extending such controls to the raising of silk cocoons and the rose harvest. Rose oil was an increasingly valuable export. By the mid-1930s, the agency was using the Agricultural Bank and the co-operative network to purchase large parts of the cotton, hemp and tobacco harvests too. The efforts of the short-lived Zveno regime to set up domestic purchasing monopolies for both tobacco and plum brandy were unsuccessful. Rising world prices after 1935 virtually eliminated grain purchases by Hranoiznos until the SecondWorld War. The agency's control of crop marketing had none the less established a precedent for close regulation of private trade. In the process, the number of private traders declined by 15 per cent between 1929 and 1939. The share of domestic trade, including non-agricultural goods, conducted by the Agricultural Bank or the co-operative network more than doubled to reach 20 per cent.[6]

The co-operative network was also responsible for a two-thirds increase in the supply of agricultural credit during the 1930s. Its rural membership of 199,000 in 1928 grew to 341,000 by 1939 — from one-quarter to one-third of the active male labour force in agriculture. Assets per member increased their lead in south-eastern Europe when combined with those of the Agricultural Bank. Marketing co-operatives accounted for almost one-fifth of the membership total. They had been joined in the 1920s by some 50 producers' co-operatives. Here were prototypes for Soviet-style collective farms without, of course, tractor stations, compulsory membership and low delivery prices. These marketing and producers' co-operatives tripled their assets from 1929 to 1939. They and the much more numerous credit co-operatives were the beneficiaries of loans from the Agricultural and Central Co-operative Banks. Loan value doubled over the same period.[7] The state budget also channelled its increased expenditures on agricultural modernisation, now up from 3 per cent to 5 per cent of total outlays, through the co-operative network.

The price for all this was paid by individual peasant households. Their access to short and long-term credit from the two banks, or from one after their merger in 1934, declined in direct proportion to the increase in credit to co-operatives. The individual peasants' share of the loans fell from 70 per cent to 30 per cent. Relief from unpaid individual debts was slow in coming any other way. The Mushanov regime forgave up to 50 per cent of unpaid debt in 1932, but only if creditors agreed. The Zveno regime offered less conditional relief, but typically placed most of its hopes for agricultural assistance on streamlining the credit system by merging the Agricultural and Central Co-operative Banks. The operations of the new bank did elimate some duplication and inefficiency, but also facilitated central control from the bank's headquarters in Sofia. The royal regimes that followed still felt obliged to grant additional relief on unpaid individual debts.[8] Yet the private, non-coop share of Bulgarian agricultural credit continued to drop. By the late 1930s it had fallen to below 40 per cent, compared with 75 per cent for neighbouring Yugoslavia and Romania.

Several modernising trends began under co-operative auspices, but the Second World War intervened before they had a chance to develop very far. The most badly needed was the consolidation of parcelled strips on smallholdings. Their size continued to decline in the absence of primogeniture. By 1934, properties under 5 hectares constituted 30 per cent of all private cropland and those under 10 hectares 69 per cent, compared with 24 and 53 per cent in 1908. Some consolidation finally got under way in the late 1930s. So did the revived importation of steam-powered agricultural equipment and the spread of irrigation typically needed for crop diversification. The dairy production that had been at the centre of Danish economic development also began to spread. The crucial change was wider availability of modern equipment for processing. Its purchase became feasible not only for co-operatives, but also for individual households, who took advantage of new mortgage regulations which accepted animals, not just land, as collateral. Co-operative credit plus specific promotion from state agronomists encouraged cultivation of higher-quality cotton. Enough was planted to increase domestic fibre and yarn production ninefold between 1929 and 1939. Total weight rose from one-quarter to almost twice the imported tonnage.[9] Here was the most immediate, if least permanent with the start of Soviet imports (see Chapters 4 and 5), of these promising trends.

Agricultural Recovery

The record of Bulgarian agricultural performance during the 1930s suggests that co-operative credit and other initiatives promoted growth more than Hranoiznos did. Grain exports were only 8–13 per cent of total export value by this time. The real output of grain per capita did increase, as noted in Table 3.1, but by about the same percentage over the output in 1926–30 in both 1931–5 and 1936–40. Purchases by Hranoiznos predominated in the first half of the 1930s, but not in the second. In the later period, moreover, rising output per hectare explained most of the increase. Wheat yields for 1934–9 were fully 30 per cent greater than the 1907–11 average. Earlier in the 1930s, expanding hectarage had explained more of the increased output than rising yields.

The most striking agricultural growth, in any case, occurred outside grain cultivation. By 1936–8 industrial crop hectarage — increasingly cotton and especially sunflowers, as well as tobacco — jumped to 234 per cent of its 1926–30 level. Output per hectare, however, fell by one-half with this expansion. The share of industrial crops in real crop output fell slightly to 15 per cent. Vegetable

Table 3.1: Indices of Crop Output, Area and Yield, 1926–38 (1926–30 = 100)

	Grain	Vegetables	Industrial Crops	Vineyards, Fruits	All crops
Gross output per capita					
1931–5	109	144	95	124	109
1936–8	118	175	107	130	122
Land in cultivation					
1931–5	106	127	127	117	109
1936–8	107	154	234	150	116
Yield per hectare					
1931–5	109	117	80	114	108
1936–8	123	127	51	102	116
Producers' prices					
1931–5	47	47	60	63	51
1936–8	62	73	84	68	66

Source: J.R. Lampe and M.R. Jackson, *Balkan Economic History, 1550–1950* (Bloomington, Indiana: Indiana University Press, 1982), table 12.2, pp. 438–41.

and vine crops gained both at their expense and at that of cereals, which slipped from 69.4 per cent to 67.2 per cent. Vegetables and vineyards rose from 13.8 per cent to 17.7 per cent. With both hectarage and yields showing sizeable gains, the real per-capita output of vegetables was 75 per cent higher in 1936–8 than in 1926–30. This increment, plus one of 38 per cent for vineyards, was primarily responsible for the overall increase of 22 per cent for all crops noted in Table 3.1. Less reliable data suggest that the per-capita value of animal products rose by 9 per cent over the same period.[10]

These real increases have been hidden from most earlier analyses by the substantially lower level of agricultural prices throughout the 1930s. Even by 1936–8, producer prices for all Bulgarian crops still stood at 66 per cent of the 1926–30 level. Vegetable and industrial crop prices had climbed back to only 73 per cent and 84 per cent of their respective levels before the depression. This was a modest recovery from a 1931–5 level which stood at just 51 per cent for overall crop prices. It was still sufficient to close completely the 20–30 per cent scissors that had opened between the prices of agricultural goods and those of manufactures throughout the period 1931–6.[11] Peasant purchasing power had, in other words, regained its 1929 level by 1938–9.

The reasons for real agricultural growth and greater crop

diversification go beyond co-operative credit and state initiative, both of which continued to display limitations. Co-operative organisations typically failed to co-ordinate their efforts even to the district level. State agronomists were assigned to district centres and did not go to the villages, except for brief visits.[12] To understand the record of real per-capita growth during the depression decade, we must also acknowledge two other influences — reduced population growth and the relatively greater efficiency of smaller peasant properties.

The peasant-led reduction in the Bulgarian birth rate, which began after the First World War, accelerated during the 1930s. According to Table 2.4, the decline was so sharp as to overcome a falling death rate and to trim the rate of natural increase to under 10 per 1,000 by 1936–40, the lowest in south-eastern Europe. A higher rate would of course have reduced the growth of per-capita production.

What rural population growth did occur was sufficient, as noted above, to subdivide peasant smallholdings still further. Total cultivated land rose by only 4 per cent from 1929 to 1939, and mainly because fallow declined from 11.7 per cent to 8.1 per cent of the arable total. As a result, the share of smallholdings under 5 hectares increased from 23.6 per cent to 30 per cent of private hectarage between the 1926 and 1934 censuses, and holdings of 5–10 hectares rose from 34.5 per cent to 36.5 per cent.

Yet, the presumption of much inter-war Western analysis that these smaller holdings were less efficient than those over 10 hectares cannot bear the weight of calculations by the contemporary Bulgarian economist A. Iu. Totev. His painstaking use of 1934 census data reveals that the net output per hectare of holdings under 5 hectares was slightly greater than for larger holdings. This greater output resulted from the application of more capital (machinery as well as animals, buildings and other inventory) and especially labour per capita. Could not a case still be made for that favourite inter-war indictment of Eastern European agriculture, 'rural over-population'? Not according to the further calculations of returns per unit of factor input (labour included) divided by the value of assets, as recently made by the American economist Marvin Jackson.[13] He finds that these rates of return were remarkably similar for holdings under 5 hectares and over 10 hectares, respectively 10 per cent and 10.2 per cent. When we subtract the greater weight of residential housing in assets of smaller holdings, they appear to have been significantly more efficient than the larger.

Such a conclusion requires two qualifications. First, smaller holdings tended to be closer to urban areas, where the incentives for diversification into the more profitable industrial and vegetable crops were greater. Second, the larger holdings were not in fact that much larger, nor their operations significantly more mechanised to permit major economies of scale or technology. Large holdings still had about 80 per cent of their assets in land and buildings, and expended only 5 per cent of their gross income on mechanised equipment, according to a survey taken by the Agricultural Faculty of Sofia University.[14]

Bilateral Trade in the German Orbit

However else we might explain Bulgaria's relatively better agricultural recovery and also its greater turn toward labour-intensive crops, the further rise of trade relations with Germany was not the strong stimulus many observers assumed it to be at the time. These relations grew from economic adversity rather than advantage. They became entangled politically with Bulgarian rearmament only during the late 1930s. Until then the Bulgarian side pursued these bilateral relations under a clearing agreement, not fundamentally different from the post-1944 arrangement with the Soviet Union. In this earlier case, as perhaps in the later one, there was no other major customer for increased exports.

Alternative markets for agricultural exports were hard to find during the depression decade. Despite accelerating industrialisation, the Soviet Union was not yet a likely customer. Stalin's economic strategy was too autarkic; Soviet foreign trade fell to 1 per cent of its national income by the mid-1930s. Further, the royal Bulgarian regime's foreign policy was too wary of expanding relations with the USSR. The commercial agreement signed between the two states under the Zveno regime was to remain a dead letter for the rest of the decade.

The Western European market for Bulgarian grain had declined during the 1920s from its pre-war peak. Access now became harder still. Bulgaria lacked even the agreements for preferential access to what remained of this Western market. Hungary, Romania and Yugoslavia had negotiated such agreements by 1932. The legacy of being on the losing side in the First World War continued to plague Bulgaria, if not Hungary.

The remaining alternative was some sort of agreement between the several Balkan states and Czechoslovakia, Central Europe's major industrial exporter after Germany. Bulgarian representatives did attend a series of conferences which began in Warsaw in 1930. The meetings explored the possibilities for regional co-operation in Eastern Europe, but little was accomplished. When Western European governments scuttled the Austro-German customs union in 1931, the prospects for any similar arrangement involving Czechoslovakia sank too. France raised political objections over any concession to Germany. The Western side also complained that a customs union would violate the principle of most-favoured nation treatment. The notion of a specifically Balkan customs union was discussed at ministerial meetings in Athens in 1930. It foundered on Greek fears of exaggerating an already large import surplus. In any case, the four states of south-eastern Europe conducted only 9 per cent of their foreign trade with each other. Their exports were too similar to expect a much greater percentage. The Balkan Tobacco Office, which they agreed to open in 1932, confined operations to futile attempts at raising world prices through output reductions. The Balkan share of world production was too small to affect prices in a glutted market.[15]

By this time, all the Central and south-eastern European countries were seeking an alternative to reduced Western European markets in bilateral clearing agreements with each other. The new agreements paid exporters and collected bills due from importers only through deposits of domestic currency in central bank accounts. Currency changed hands at a negotiated rate of exchange and only when cumulative balances were settled. This was infrequent and open to delaying tactics by the deficit country. Barter trade and systems of import licensing soon followed to trim remaining imbalances. The rapid Bulgarian turn to clearing agreements emerges clearly from Table 3.2.

For Bulgaria, exchange controls already pointed the way to bilateral clearing trade with Germany. Despite good harvests and export surpluses in 1930–1, deficits in the government budget and the balance of payments had created a classic 'transfer problem' — how to find the foreign currency needed to service the foreign debt. The Bulgarian government faced demands not only for payment on pre-1914 borrowings from largely French bondholders, but also from the Greek government. Athens was owed three-quarters of the Bulgarian reparations due in 1932. That June the Lausanne conference ratified the end of all reparations payments. The French bank Paris Bas was still threatening to stop Bulgarian tobacco production until bond-holders guaranteed their interest payments. It was therefore imperative that the Bulgarian import surplus with Western Europe be reversed as soon as possible. That reversal might offer some relief for the Bulgarian leva from pressures to devalue and thereby cut capacity for debt servicing.[16]

At this point the semi-official German trade organisation Reemstma came on the scene; Germany was just as eager as Bulgaria to conduct foreign trade that would not add to debts owed in Western Europe. Its informal clearing agreement with Bulgaria in 1931 was formalised the following year at the 1929 exchange rate of 33 leva for 1 Reichsmark, reflecting their almost equal depreciation on the free market. The German share of Bulgarian exports jumped back to 36 per cent, after declining from 42 per cent to 30 per cent between 1929 and 1931. This and other agreements had already channelled 76 per cent of Bulgarian exports, according to Table 3.2, into the clearing trade for 1932 and 1934-5. The clearing proportion rose to 88 per cent by 1939. The German share of all Bulgarian exports reached 68 per cent. As indicated in Table 3.3., however, the rise in the German and Austrian share of Bulgarian exports was not striking until 1938. The import share, on the other hand, had already doubled between 1931 and 1935.

Bulgarian reluctance to accept overdependence on a single trading partner surfaced soon after Hitler's Finance Minister, Hjalmar Schacht, visited Sofia in June 1936 to negotiate a new clearing agreement. Bulgaria thereupon sold enough wheat to Great Britain in 1937 to account for 14 per cent of total export value. Barter arrangements with Poland and Czechoslovakia the same year, the latter trading railway equipment for tobacco, served the same purpose. German deliveries of industrial raw materials had already proved disappointing. As late as January 1938, the Bulgarian War Minister was trying to meet at least a fraction of Bulgarian rearmament imports through similar barter deals with France, Poland and Italy, but to no avail. Only the German side offered arms on long-term credit. The prospect of a 30 million Reichsmark credit in March 1938 combined with the German annexation of Austria to settle the issue.[17]

Did this bilateral trade provide an export market, which helps

90 Isolation in the 1930s

Table 3.2: Foreign Trade Balances, 1926–38 (annual leva average per capita in 1929 prices)

	Exports	Imports	Index Exports	Index Imports	Percentage clearing Exports	Percentage clearing Imports
1906–10			68	78		
1926–30	1,414	1,177	100	100		
1931–5	1,281	693	90	58	76[a]	76[a]
1936–8	1,381	876	97	74	81	89

[a] 1932, 1934–5.
Source: Lampe and Jackson, *Balkan Economic History, 1550–1950*, table 12.8, pp. 462–3.

Table 3.3: Direction of Foreign Trade, 1929–39 (percentage)

	North-western Europe[a]		Germany and Austria		Central and North-eastern Europe[b]		South-eastern Europe and Turkey	
	X	M	X	M	X	M	X	M
1929–31	19	28	41	30	17	13	7	11
1932–4	23	26	45	40	7	9	2	7
1935–7	22	14	50	62	10	12	2	5
1938–9	11	12	63	59	9	13	2	5

X = exports; M = imports.
[a] France, Holland, Belgium, Great Britain, Switzerland and Italy.
[b] Hungary, Czechoslovakia, Poland and USSR.

Source: Lampe and Jackson, *Balkan Economic History, 1550–1950*, table 12.7, pp. 458–60.

to explain the relatively better Bulgarian agricultural performance during the 1930s? The value of German purchases of Bulgarian exports, virtually all agricultural, did double from 1930 to 1938, after a slow start. Close inspection of their composition none the less reveals a pattern of growth that did not favour the new labour-intensive crops, let alone their further processing. Tobacco sales doubled in value, and rose 3.5 times by weight, to account for 41 per cent of German purchases by 1938. Eggs dropped from 31 per cent to 13 per cent, but were replaced by table grapes, up from nil to 16 per cent. Exports of more labour-intensive crops like tomatoes got under way, but their totals were minimal. Tomatoes accounted for less than 1 per cent of 1938 exports to Germany (and Austria). Nor did new processed exports to Germany advance

Isolation in the 1930s 91

noticeably. Some 48 canning plants, mainly in the Plovdiv area, were pulping tomatoes and strawberries by 1939. Yet pulp sales were just 1.5 per cent of Bulgarian exports to the Reich. For strawberry pulp, sales to Great Britain were actually larger, despite the absence of a clearing agreement and a favourable exchange or tariff rate.[18]

By 1936–8, according to Table 3.2, Bulgaria's real exports per capita had admittedly climbed back to 97 per cent of their 1926–30 level. Yet this was no better than the world average or Yugoslavia's performance, and worse than Romania's. The latter's 122 per cent had been achieved mainly through oil exports to Western markets. The German share in Romanian exports was less than one-quarter. When we include the unsatisfactory record of Bulgarian imports from Germany, it is therefore tempting to extend Philip Friedman's negative appraisal of the effect of German bilateral trade on Hungary to this small Balkan economy.[19] Bulgarian agricultural performance, particularly the 22 per cent increase in crop value, remains impressive, but might have been even better in a wider, multilateral world market. Between 1929 and 1939, foreign-trade turnover had after all declined from nearly 40 per cent to less than 30 per cent of our notion of national income.

The Predominance of State Banks

More clearly, the conversion of so much foreign trade to bilateral agreements aided the ascendancy of what was already the strongest state sector among the banking systems of south-eastern Europe. The central bank saw its assets regain their 1929 level by 1937, despite the region's most restrictive policy of note issue.[20] This policy kept the leva within a few points of its 1929 par value with gold throughout the 1930s. Restriction perpetuated the overvalued free exchange rate of the previous decade and discouraged any Bulgarian exports outside special clearing or compensation agreements. As was customary elsewhere, the central bank became the administrator of these agreements. The Bulgarska Narodna Banka thereby acquired new powers. They more than made up for the loss of commercial lending rights in 1928. From January 1933, the bank's authorisation was required for all imports. By 1936, the Narodna Banka was supervising the aforementioned system of import licences and quotas which the clearing agreements had fostered.

It also monitored the export trade by means of regulations for 'private compensation', under which Bulgarian exporters might sell their earnings of foreign currency to importers of goods, primarily from Western Europe, outside the clearing agreements. Such compensation arrangements covered 35 per cent of Bulgarian foreign trade by 1938, including some under the German clearing agreements. This complex system included premiums up to 35 per cent beyond the official exchange rate, charged for leva conversion of Western currency earned for exports of grain, eggs, tobacco and other goods considered to be glutting the European market. Dairy products paid lower premiums and fruit or vegetables none at all.[21] In this way the Narodna Banka hoped to encourage the export of the new labour-intensive crops and made sales, even to Great Britain, as we have seen, more attractive to exporters. The bank assigned these premium-free conversions to badly needed imports like industrial raw materials, shunting those for grain, etc. into full-premium transactions for luxury imports.

While the central bank was consolidating new powers not reflected in its balance sheets, the Bulgarian Agricultural Bank (BZB) was adding to the assets which had already made it the country's largest financial institution by the 1920s. Those assets doubled again between 1929 and 1939. Deposits in 1939 had risen by 42 per cent since 1931, buoyed in 1934 by a merger in that year with the state's small Central Co-operative Bank. This project of the Zveno regime had added only 14 per cent to deposits, but had never the less strengthened the authority of the BZB. Between 1931 and 1939, its share of new bank credit had climbed impressively from one-third to one-half.

Some 19 native commercial banks were also assembled to form a new state institution, the Banka Bulgarski Kredit. The Zveno regime had created it in 1934, with the aim of putting all of the country's private banks outside Sofia under one roof. Despite over twenty failures since the start of the depression, they still numbered nearly a hundred. The new state bank did not come close to absorbing them all, but tended to attract the largest ones. Its commercial credit, when combined with that of the Agricultural Bank, accounted for three-quarters of the new bank credit granted between 1936 and 1939. This state system also included the Popular Banks, which were loosely affiliated with the co-operative network. Their savings deposits rose by one-third during the 1930s. Deposits in the state system thereby jumped to two-thirds of the Bulgarian total by 1937. The state share of bank assets approached three-quarters.[22]

Private commercial banks never recovered from the initial shock of the depression. Their assets had declined by one-half between 1929 and 1934, and turned up only slightly thereafter. Foreign banks led the retreat. Their share of commercial bank assets fell from 39 per cent to 21 per cent (from 21 per cent to only 7 per cent, if state assets are included) between 1931 and 1939. First for economic, and then also for political reasons, Franco-Belgian interests largely abandoned their joint-stock shares in the large Sofia banks which had lent and invested so widely during the 1920s. The state bought out their Sofia Mortgage Bank in 1936. The large Italian bank simply held its ground. Germany's Kreditna Banka increased its assets and activities, but hardly enough to make up for the Western European withdrawal.[23]

As a result, bank investment in Bulgarian industry also dropped precipitously. The European bank share in firms with some foreign capital fell from four-fifths to one-fifth. By 1939 the German Kreditna Banka accounted for almost half of that small fraction. The Bulgarian Agricultural Bank had meanwhile bought out the French and Czech ownership in the country's half-dozen modern sugar refineries. Their private cartel for controlling production and prices had already been replaced at the start of the decade, it may be recalled, by state regulation.

Industrial Growth without Concentration

The general effect of the depression on domestic demand and the specific loss of Western European bank investment obviously limited the prospects for Bulgarian industrial growth during the decade. At the same time, an authoritative British survey of south-eastern Europe could already conclude in 1936 that the depression had transformed industrialisation 'from the political desideratum which it had largely been in the previous period [the 1920s] into a vital economic necessity'.[24] Falling agricultural exports limited manufactured imports, thus attracting domestic industry to import substitution. More manufacturing would also increase the potential for processed exports.

Despite some serious limitations, Bulgarian industry at least overcame the general European pattern of stagnation or decline.

Isolation in the 1930s

Real output increased by 52 per cent between 1929 and 1938, at an average annual rate of 4.8 per cent. This rate was well beyond the European average of 1.1 per cent, and ahead of the 3.4 per cent for neighbouring Yugoslavia and Romania. Only Swedish and Finnish rates matched the Bulgarian one, and only the Greek and the Soviet rates exceeded it. Recent Bulgarian calculations record an average industrial growth rate of 6.9 per cent for the latter period of 1935–41. One-third of all processed foodstuffs were exported by 1939. Yet manufactured goods still amounted to only 4.3 per cent of export value, down from 8.1 per cent in 1929, because of the decline in textile sales.[25]

Import substitution was responsible for most of Bulgaria's industrial growth during the 1930s. The domestically produced share of industrial output consumed in the country rose significantly, from 61 per cent to 88 per cent between 1929 and 1938. The imported share of industrial inputs, that is new materials and semi-finished goods, continued to be high. Imports still represented 38 per cent of industrial input value in 1938, as against 40 per cent in 1930. Yet the structure was changing. Metals and machinery accounted for half of the value of all imports by 1938. After 1932, moreover, the imports of cotton yarn fell from three-quarters to one-third of domestic consumption, while cotton cloth production increased.[26] Here were semi-finished goods whose earlier import Alexander Gerschenkron had called a classic example of Bulgarian industrial (and agricultural) immaturity. By the mid-1930s domestically produced yarn, increasingly spun from home-grown fibre, was replacing imports of both.

Modern industry none the less made little progress increasing the share of Bulgarian national income during the 1930s. A recent American calculation of Bulgarian national income for 1938 by Marvin Jackson finds modern industry accounting for just 5.6 per cent of the total, by far the lowest fraction in south-eastern Europe. This represents only a slight increase over the 5.1 per cent reckoned for 1926, under admittedly different assumptions, by the Bulgarian pioneer in national income accounts, Asen Chakalov.[27] Still more disturbing, if we dare call these two sets of calculations comparable, is the *increase* from 1926 to 1938 in the share of artisan production, from 5 per cent to 9.3 per cent. Rural domestic industry declined only slightly, from 9.1 per cent to 7.3 per cent. Mechanised manufacture in firms of ten or more employees was thus losing a little ground to pre-modern enterprises during the depression decade. The former's share of the labour force did rise, but only slightly, from 13 per cent to 14 per cent.

Even among those Bulgarian manufacturers meeting this modest definition of a modern enterprise, the smallest accounted for most of the decade's industrial growth. The larger, typically joint-stock enterprises, which Bulgarian and Western economists both find essential for technological innovation and economies of scale, were almost entirely covered by the encouragement legislation for tax and tariff exemptions. For reasons that the next section will make clear, most of these encouraged enterprises hardly grew at all during the 1930s. Table 3.4 reveals that the increasingly numerous unencouraged factories were almost entirely responsible for industrial production (the stagnant and unencouraged tobacco branch aside) growing at an annual average of 6.4 per cent from 1931 to 1937. During those years, according to Table 3.5, the number of encouraged enterprises fell from 1,145 to 854. The unencouraged rose from 643 to 2,031.

Output of processed food recorded the highest growth rate, except for the small paper branch. The number of food enterprises grew even faster, doubling during this period. Hard times in agriculture undoubtedly diverted some peasants and traders from selling their produce to processing it. Virtually all of the new firms were unencouraged. Similar trends in most other branches of manufacturing kept the average size of Bulgarian manufacturing enterprises the smallest in south-eastern Europe. With unencouraged firms included, that size fell from 32.5 to 28.6 employees for 1931–8. (Table 3.5 calls our attention to the larger size of encouraged firms.) The average annual growth of labour inputs and horsepower for Bulgarian industry between 1927 and 1937 increased by just 2 per cent and 3.2 per cent, compared with 4.2 per cent and 4.8 per cent for Romania.[28] Although the number of joint-stock enterprises did not stop rising during the decade, those established from 1931 to 1937 accounted for less than one-fifth of the new industrial firms.

What growth was occurring during the depression does not therefore generally fit the model for increased concentration and monopoly profits to which Marxist scholars have typically looked to explain the rise of industrial capitalism. Larger firms with over 50 employees did continue to increase in number during the early 1930s. After 1934, however, growth in numbers and output was coming instead from a flood of new small-scale enterprises, many

Table 3.4: Growth of Net Industrial Output, 1921–37 (average annual percentage)

	All enterprises		Encouraged enterprises	
	1921–31	1931–7	1921–31	1931–7
Metals and machinery	23.5	1.8	26.1	−0.3
Chemicals	27.0	5.9	28.8	1.8
Construction materials	10.4	9.8	13.7	3.4
Wood processing	7.2	0.3	17.2	−6.7
Textiles	23.5	3.5	24.4	2.1
Leather	9.7	6.1	12.2	10.1
Tobacco	−0.9	1.0		
Flour	6.7	−12.2	11.9	−12.9
Other food	8.1	13.8	7.9	−1.9
Total, excluding tobacco	13.3	6.4	15.6	0.3

Source: M.R. Jackson and J.R. Lampe, 'The Evidence of Industrial Growth in Southeastern Europe before the Second World War', *East European Quarterly*, vol. XVI, no. 4 (1983), p. 396.

Table 3.5: Number and Size of Industrial Firms in 1937

	All enterprises		Encouraged enterprises	
	No.	Average no. of workers	No.	Average no. of workers
Metals and machinery	131	43	103	45
Chemicals	271	15	96	27
Construction materials	141	35	87	44
Wood processing	142	19	46	25
Paper	15	112	4	243
Textiles	382	82	197	125
Leather	63	22	46	19
Food and tobacco	1,740	22	275	19
Total	2,885	31	854	52

Source: Jackson and Lampe, 'Evidence of Industrial Growth in Southeastern Europe', table 12.18, pp. 393–4.

of whose owners had doubtless been tied to agricultural exports. Most Western economists would, however, join their Bulgarian colleagues in finding such growth an unlikely vehicle for modern industrial development.

Sofia as an Industrial Centre

Sofia was the main centre of Bulgaria's unencouraged industrial growth during the 1930s, as it had been for state-encouraged growth before and after the First World War. One-half of Sofia's industrial enterprises, typically small and unencouraged, had been founded between 1931 and 1938. By then the capital city accounted for one-third of all enterprises and one-quarter of their production. The city's proportion of the labour force in modern industry was even higher, over half of the 100,000 estimated by 1939.[29]

Here was a kind of concentration, if not in the classic Marxist mold of a shrinking number of even larger enterprises. The reasons for this geographic concentration of new and expanding, though small firms, lay mainly on the demand side. Sofia's population had continued to rise during the 1930s, in part because of job seekers leaving a depressed rural economy. Sofia had grown to 287,000 by 1934, with another 41,000 in nearby villages. Its share of Bulgaria's urban population had jumped from 16 per cent in 1920 to 22 per cent by 1934. The city's modern mass culture included foreign films and native radio. These media joined with the greater availability and advertising of consumer goods to provide a positive attraction. Sofia was the country's main entrepôt for imported consumer goods. Their flow into Sofia had increased during the depression decade. Their value was still double that of manufactures produced in the city by 1937.[30] Although priced high because of tariff impositions, these imports still sold out and increased consumer demand through the so-called demonstration effect. They also encouraged domestic manufacturers to undercut them. Import substitution was encouraging industrial growth in Bulgaria most successfully, in other words, where import competition was still the greatest.

The supply side admittedly made some contribution to Sofia's industrial growth as well, under conditions which Bulgarian Marxist scholars have rightly criticised. Industrial employment increased by perhaps one-half between 1934 and 1939, but without any significant increase in wages relative to other Bulgarian towns. A continuing inflow of new job seekers kept unemployment at one-quarter of the available labour force. The unemployed and most of the employed subsisted in unauthorised shacks, jerry-built on the city's western outskirts, together with the Macedonian refugees of the 1920s. Hygienic conditions were bad, and in the few factory barracks even worse.[31] Those employed in industry none the less constituted 30 per cent of the city's active labour force, far surpassing the state employees who had outnumbered them 3 to 1 before the First World War.

Supplies other than labour were less favourable for Sofia's private manufacturers. We have already seen the contraction in commercial lending from private banks. In addition, the city's industrial enterprises faced a scarce supply of coal and an expensive supply of electricity. The nearby Pernik mine at least favoured Sofia over other Bulgarian towns, but the first claim of the state railway on its coal continued to plague private owners from time to time. The city's foreign-owned electric power station posed more of a problem. Again, because of Bulgaria's alliance with the losing side in the First World War, the electric system's Franco-Belgian pre-war concessionaires had been able to regain control by 1926 under League of Nations' arbitration. They raised prices to recapture lost profits, but added little new capacity. They also used the terms of their concession to prevent several private manufacturers from building their own power plants.[32] Such conditions discouraged private use of the major technological advance available to inter-war European industry.

Industrial Policy and the State Sector

It was public policy rather than private monopoly power that, together with the world depression, held back encouraged industry. The virtual zero-growth for these enterprises, recorded in Table 3.4 above, testifies to the effectiveness of the restraints. As with agriculture, the main lines of state policy toward industry during the depression decade were consistent from the last year of the Liapchev regime to the royal cabinets.

The principal policy was to withdraw encouragement from most branches of light industry. The depression had already reduced production in encouraged sugar, foodstuffs, textile and leather enterprises to 50–70 per cent of capacity by 1931. That year Liapchev's Minister of Trade and Industry introduced the notion that Bulgaria's major light industries were 'saturated' with too many enterprises. Only the selective withdrawal of the tax and tariff exemptions accorded in the industrial law of 1928, it was argued, would relieve this imbalance. The new Malinov-Mushanov regime of the National Bloc initially opposed such a policy. Its supporters included owners of many small-scale enterprises which either made or sold encouraged manufactures. Already, high import tariffs on finished goods were boosted in 1931 to *ad valorem* rates that averaged 90 per cent. Within less than one year, however, a Bulgarian consensus to accept the saturation argument asserted itself. The Mushanov regime began quietly withdrawing tax and tariff exemptions from existing firms and tightening the requirements of any new firm applying for exemption.

In 1933, it took the further steps of imposing higher tariffs on most imported industrial inputs and of limiting exemptions. Rates ranged to 60 per cent *ad valorem*. The regime's proposal to empower the Ministry of Trade and Industry to fix the prices of all industrial inputs, domestic as well as imported, was not implemented until 1936. At that time the royal regime also announced the formal withdrawal of all encouragement provisions from 17 branches of manufacturing.[33] Together they accounted for almost one-half of Bulgarian industrial production. In the words of an officially favoured Bulgarian economist, the protected production of finished consumer goods stood condemned as a 'dead-end' for Bulgarian industry.[34] More promising, he argued, would be the manufacture of industrial inputs like cellulose and textile dyes. Competition with German imports, however, proved too stiff for domestic production of these particular inputs. The major success appeared in the aforementioned rise in domestic production of cotton fibre and yarn. From 1936, the Ministry of Trade and Industry helped by setting their prices under those charged for dutied imports. Price controls were in fact the major new precedent set for state industrial policy during the 1930s.

At the start of the decade, it had appeared that state supervision of private cartels would be the principal source of new leverage. Apart from the growing influence of the Agricultural Bank in the sugar cartel, however, state use of the 1931 regulations for cartel registration remained minimal. Existing cartels simply reported their activities. The 25 new cartels formed during the 1930s typically drew their members from smaller firms in provincial towns and lasted only one or two years.[35] The flood of new enterprises in Sofia largely avoided cartel membership in order to be able to maximise their sales. Food processors were especially reluctant to join cartels. They relied strictly on domestic inputs and counted on the continuing tariff barriers to protect them against imported output.

Public policy toward industry thus resulted in *less* state control during the 1930s, rather than more as with agriculture, foreign trade and finance. The Zveno regime did its part, for efficiency, if not for decentralisation, by dismissing some 4,000 of the 90,000 state employees, and by attempting to streamline the operation of the bureaucracy. All economic ministries were consolidated into one. The royal regime's growing interest in rearmament and in enlarging Bulgaria's small army, limited by treaty to 20,000 until 1934, did not extend to plans for mobilising industrial production, either through cartel management or through direct state ownership. The largest part of the doubling in military expenditures that occurred between 1934 and 1939 went to pay for motor vehicles imported from Germany. Only domestic cement production appears to have benefited from the build-up. Although Andrei Liapchev had written in 1933 of the need for some sort of coordinated planning for the entire economy, nothing to include industry was even proposed until the Second World War.[36]

This is not to say that the state's industrial activity was negligible. The prospect of state regulation was omnipresent. Factory owners hesitated to take initiatives which they did not regard as officially favoured. In the words of one contemporary observer, 'Along with the many employees in every industrial enterprise, there is one unseen, but most important employee, something like a 51 percent shareholder — that is the state.'[37]

State-owned industry was also growing, especially if mining and infrastructure are included. By 1939 some 169 state enterprises, led by the Pernik coal mines, accounted for 8–9 per cent of total industrial production. Co-operatives produced another 5–6 per cent, typically small-scale enterprises except for several sugar refineries. This percentage had nearly doubled during the 1930s. Railway, road and public building construction increased from 1934 forward. Railway trackage rose from 2,487 km in 1929 to 3,123 km by 1939. Locomotives and rolling stock also increased enough to allow freight tonnage to rise by one-quarter. New public buildings, especially in Sofia, were being constructed although without the use of new technology. In order to provide as much public employment as possible, construction sometimes proceeded without automatic cement mixers.

The size of the state's economic activity was thus significant, despite the absence of much effort to co-ordinate or modernise it. The state and co-operative share of 1939 industrial output, about 15 per cent, now exceeded the figure for foreign investors. The latter had fallen by one-third during the decade. In addition, state revenues had increased by 60 per cent in real terms during 1929–39, a pace well ahead of the zero growth for foreign trade. Sofia's municipal revenues had tripled over the period.[38] In both cases, earnings from economic enterprises were the largest and most rapidly growing source of income. The shortfalls in tax caused by the depression were ironically the major force in expanding the production, if not the scope, of state industry during the decade.

Notes

1. The most comprehensive and sympathetic account of the Zveno regime remains that of George C. Logio, *Bulgaria Past and Present* (Manchester: Sherratt and Hughes, 1936), pp. 1–49. A sympathetic view of the National Bloc regime is that of Nissan Oren, *Revolution Administered: Agrarianism and Communism in Bulgaria* (Baltimore, Maryland: Johns Hopkins Press, 1973), pp. 5–43. Recent Bulgarian scholarship, critical of both regimes, is represented by Vladimir Migev, *Utvurdzhavane na monarkho-fashistkata diktatura v Bulgariia, 1934–1936* [The Consolidation of the Monarcho-Fascist Dictatorship in Bulgaria, 1934–1936] (Sofia, 1977).

2. Boris was never the less the first foreign leader received by Hitler in the first of their several congenial meetings. On Boris's German dealings and the Nazi's Bulgarian allies, see Hans-Joachim Hoppe, *Bulgarien: Hitlers einwilligen Verbundeter* (Stuttgart, 1979), pp. 39–46; and Georgi Markov, *Bulgarogermanskite otnosheniia. 1931–1939* [Bulgarian-German Relations, 1931–1939] (Sofia, 1984). The latter volume is an outstanding piece of Bulgarian scholarship on the 1930s. On Bulgarian right-wing ideology, see Boris Stavrov, *Buzhoaznata sotsiologiia v Bulgariia mezhdu dvete svetovni voini* [Bourgeois Sociology in Bulgaria between the Two World Wars] (Sofia, 1979), pp. 149–67.

3. Michael Kaser, 'The Depression and the Genesis of Etatism', in Vasa Čubrilović (ed.), *Svetska eckonomska kriza, 1929–1934 g.* [The World Economic Crisis, 1929–1934] (Belgrade, 1976), pp. 54–5.

4. Lynn R. Edminister, Leo J. Schaben and Myer Lynsky, *Agricultural Price-Supporting Measures in Foreign Countries*, F. S. no. 56 (Washington, DC: US Department of Agriculture, 1932), pp. 44–8. On the general experience of countries that were agricultural exporters, see Charles P. Kindleberger, *The World in Depression, 1929–1939* (Berkeley, California: University of California Press, 1973), pp. 83–107, 146–72.

5. Nikolai Todorov et al., *Stopanska istoriia na Bulgariia, 681–1981* [Economic History of Bulgaria, 681–1981] (Sofia, 1981), pp. 369–70.

6. Ibid., pp. 396–7; Edminister et al., *Agricultural Price-Supporting Measures*, pp. 49–52.

7. *Statisticheski godishnik na Bulgarskoto Tsarstvo* [Statistical Yearbook of the Bulgarian Kingdom], *1935* (Sofia, 1936), pp. 608, 618–19; *1940* (Sofia, 1940), pp. 311, 320–1.

8. League of Nations, *Chronology of Political and Economic Events in the Danubian Basin, 1918–1936: Bulgaria* (Paris, 1938), pp. 50–2. Foreclosures at least could not deprive a peasant household of the minimum 2–3 hectare holding, which Stamboliiski's regime had made inalienable after the First World War.

9. John R. Lampe and Marvin R. Jackson, *Balkan Economic History, 1550–1950: From Imperial Borderlands to Developing Nations* (Bloomington, Indiana:

Indiana University Press, 1982), p. 455, and table 12.17, pp. 489–90. On state and co-operative aid to agriculture, see J.S. Moloff, 'Bulgarian Agriculture', in O.S. Morgan (ed.), *Agricultural Systems of Middle Europe* (1933; reprinted, New York: AMS Press, 1969), pp. 62–75. The distribution of private land is summarised in table 31 of Nikolai Todorov *et al.*, *Stopanska istoriia na Bulgariia*, p. 373.

10. See Lampe and Jackson, *Balkan Economic History*, table 12.4, pp. 447–8; Marvin R. Jackson, 'Agricultural Output in Southeastern Europe, 1910–1938', *ACES Bulletin*, vol. XXIV, no. 4 (1982), pp. 51–6, 63.

11. Lampe and Jackson, *Balkan Economic History*, table 12.2, pp. 438–41; table 29 in Todorov *et al.*, *Stopanska istoriia na Bulgariia*, p. 371.

12. The limits emerge from the best case-study of an individual village during the 1930s, Irwin T. Sanders, *Balkan Village* (Lexington, Kentucky: University of Kentucky Press, 1949) on the social anthropology of Dragalevtsi, near Sofia.

13. Lampe and Jackson, *Balkan Economic History*, table 12.3, pp. 442–5. The pioneering Bulgarian work is A. Iu. Totev, *Sravnitelno izuchavane na bulgarskoto-iugoslovenskoto narodno stopanstvo* [A Comparative Inquiry in the Bulgarian and Yugoslav Economies] (Sofia, 1940).

14. Totev, *Sravnitelno izuchavane*, pp. 52–61; Todorov *et al.*, *Stopanska istoriia na Bulgariia*, pp. 368–76.

15. Theodore I. Geshkoff, *Balkan Union* (New York: Columbia University Press, 1940), pp. 145–61; Edminister *et al.*, *Agricultural Price-Supporting Measures*, pp. 52–4.

16. This was in fact not accomplished until 1936, according to Table 3.3. The debt-service ratio had risen to 12.7 per cent by 1930 because of recent loans; peaked at 14.3 per cent in 1931–2; and then fell to 5 per cent by 1935. The debt's share of budget expenditures declined from 32 per cent in 1931–2 to 20 per cent by 1939. See Lampe and Jackson, *Balkan Economic History*, table 12.13, pp. 480–1, 457–64. On the Paris Bas dispute and its subsequent history, see William H. Wynne, *State Insolvency and Foreign Bondholders* vol. II (New Haven, Connecticut: Yale University Press, 1951), pp. 560–73.

17. Hoppe, *Bulgarien*, pp. 46–9; Lampe and Jackson, *Balkan Economic History*, table 12.8, pp. 462–3.

18. *Statisticheski godishnik na Bulgarskoto Tsarstvo, 1935*, pp. 256–9; *1940*, pp. 533–42. Bulgarian exporters outside the clearing agreements paid up to a 35 per cent premium for access to the earnings in leva still formally pegged to gold, an effective devaluation of 26.5 per cent. On the German clearing trade, see Howard S. Ellis, *Exchange Control in Central Europe* (Cambridge, Massachusetts: Harvard University Press, 1941), pp. 259–89.

19. Philip Friedman, 'The Welfare Costs of Bilateralism: German-Hungarian Trade, 1933–1938', *Explorations in Economic History*, vol. 13, no. 1 (1976), pp. 113–25. On Romania, see Phillipe Marguerat, *Le III Reich et le petrol roumain, 1938–1940* (Leiden, 1977).

20. See Lampe and Jackson, *Balkan Economic History*, tables 12.10–12.12, pp. 471–7.

21. S.D. Zagoroff, J. Vegh and A.D. Bilimovich, *The Agrarian Economy of the Danubian Countries, 1933–1945* (Stanford, California: Stanford University Press, 1955), pp. 375–6.

22. Lampe and Jackson, *Balkan Economic History*, table 12.2, pp. 476–7. The best-running account of Bulgarian financial and budgetary affairs during the 1930s is in the annual publication, Overseas Trade Department, Great Britain, *Economic Conditions in Bulgaria* (London: HM Printing Office).

23. Mirko Lamer, 'Wandlungen der ausländischen Kapital auf der Balkan', *Weltwirtschaftliches Archiv*, vol. 48 (1938), pp. 501–17; Todorov *et al.*, *Stopanska istoriia na Bulgariia*, pp. 360–5, 399–400. Bulgarian Marxist scholarship has long

since abandoned its 1950s argument that French and Belgian capital left Bulgaria in the 1930s as part of a capitalist conspiracy to make way for German investment. In fact, many of the Western investors were Jewish and left largely out of fear.

24. Royal Institute of International Affairs, *The Balkan States*, vol. I: *Economic* (London: Oxford University Press, 1936), p. 115.

25. *Statisticheski godishnik na Bulgarskoto Tsarstvo, 1940* p. 558. On industrial growth rates, see Marvin R. Jackson and John R. Lampe, 'The Evidence of Industrial Growth in Southeastern Europe before the Second World War', *East European Quarterly*, vol. XVI, no. 4 (1983), pp. 392–8.

26. Lampe and Jackson, *Balkan Economic History*, tables 12.16–12.17, pp. 485–8.

27. Agriculture was still the leading sector, with crops accounting for 30 per cent of 1938 national income, animals 13.8 per cent and trade (largely agricultural) 16.5 per cent. See Marvin R. Jackson, 'National Product and Income in Southeastern Europe before the Second World War', *ACES Bulletin*, vol. XXIV, no. 3 (1982), table 2, pp. 80–1. For a recent Bulgarian summary of previous estimates, see Petur Shapkarev, *Statistiko-ikonomicheski etiudi vurkhu narodnoto stopanstvo na NR Bulgariia* [Statistical-Economic Studies on the National Economy of the PR of Bulgaria] (Varna, 1982), pp. 121–6.

28. Lampe and Jackson, *Balkan Economic History*, tables 12.15 and 12.18, pp. 485, 493–5; Todorov *et al.*, *Stopanska istoriia na Bulgariia*, pp. 360, 436.

29. Khristo Marinov, 'Geografsko razpredelenie na promishlenostta v Bulgariia mezhdu dvete scetovni voini' [The Geographic Distribution of Industry in Bulgaria between the Two World Wars], *Trudove na V.I.I. Karl Marx*, vol. I (1965), pp. 7–14.

30. See John R. Lampe, 'Interwar Sofia vs the Nazi-style Garden City: The Struggle over the Muesmann Plan', *Journal of Urban History*, vol. II, no. 1 (November 1984), pp. 39–62.

31. See Liuben Berov, *Polozhenieto na rabotnicheska klasa v Bulgariia pri kapitalizm* [The Position of the Working Class in Bulgaria under Capitalism] (Sofia, 1968), pp. 25–100. State health and unemployment insurance had been introduced for urban workers in 1924–5, and expanded in 1937, but its provisions covered only a fraction of expenses. L. Radulov (ed.), *Sotsialno-ikonomicheskata politika na Bulgarskata durzhava, 681–1981* [Socio-economic Policy of the Bulgarian State, 681–1981] (Varna, 1981), pp. 181–3.

32. Sofia's electric capacity per capita was one-third of the levels for Belgrade and Bucharest during the 1930s. Lampe, 'Interwar Sofia', pp. 53–4; and Liuben Berov, 'Kontsesiiata za elektricheskoto osvetlenie v Sofiia, 1900–1941' [The Concession for Electric Lighting in Sofia], *Izvesttiia na Instituta za istoriia*, vol. 12 (1963), pp. 75–109.

33. Overseas Trade Department, *Economic Conditions in Bulgaria, 1934*, pp. 36–40; *1937*, pp. 14–18; Todorov *et al.*, *Stopanska istoriia na Bulgariia*, p. 377–84. On tariff levels, see Lampe and Jackson, *Balkan Economic History*, table 11.10, pp. 412–13.

34. St. Bobchev in *Spisanie na bulgarskoto ikonomichesko druzhestvo*, vol. XXXVII, no. 5 (1938), pp. 273–81.

35. Liuben Berov, 'Kum vuprosa za monopolisticheskite organizatsii v Bulgariia' [On the Question of Monopolistic Organizations in Bulgaria], *Ikonomicheska misul*, vol. 7 (1958), pp. 67–80.

36. Zhak Natan (ed.), *Istoriia na ikonomicheskata misul v Bulgariia* [The History of Economic Thought in Bulgaria], vol. 2 (Sofia, 1973), p. 174. On the growth of state employment and the army, see Lampe and Jackson, *Balkan Economic History*, table 12.22, pp. 505–8.

37. Zh. Burilkov, 'Strukturni promeni v industriiata' [Structural Changes in

Industry], *Spisanie na bulgarskoto ikonomichesko druzhestvo*, vol. XXXVII, no. 10 (1938), pp. 640–1.

38. Lampe and Jackson, *Balkan Economic History*, table 12.13, pp. 480–1; Lampe, 'Interwar Sofia', pp. 53–4. On the growth of infrastructure and state industry, see Todorov, *et al.*, *Stopanska istoriia na Bulgariia*, pp. 359–400.

4 THE SECOND WORLD WAR

In 1941, for the second time in 30 years, Bulgaria was drawn into a world war by ill-fated alliance with Germany. As in 1915, German representatives were able to play successfully on the small country's lack of Western European allies and on its frustrated claim to Macedonia. The closer economic and then military relations that had developed with Germany during the 1930s also supported the Bulgarian decision. Tsar Boris increasingly lent his personal support. In any case, by August 1939, Nazi Germany had come to diplomatic terms with the Soviet Union, and within the next year crushed Poland and France on the battlefield. After the fall of France, Bulgaria abandoned its formal neutrality between Germany and the Western allies. By September 1940, German and Soviet pressure had forced the Romanian government to cede the southern Dobrudja back to Bulgaria.

The war itself also began auspiciously for Bulgaria. The government signed the Tripartite Pact of military alliance with Germany and Italy in March 1941. The German army had already assembled over half a million men in Bulgaria for Operation Maritsa against Greece. This force was primarily responsible for the Blitzkrieg that overwhelmed Yugoslavia in April and Greece in May. The Nazi leadership invited the Bulgarian army to occupy Macedonia and Thrace. The Bulgarian government formally annexed most of this territory. One well-quoted Bulgarian observer called these first months of the war 'a lovely dream' from which he feared a 'terrible awakening', that would include the loss of territory given as a gift rather than earned.[1]

The rest of the war amply fulfilled his prophecy. Occupation duties drew the Bulgarian army further into Yugoslav and Greek territory. By 1942, its units were facing Communist-led resistance movements. The increasing burden on Bulgarian manpower and resources at least helped the government to resist repeated Nazi pressures to join the German war on the Soviet Union. The Bulgarian government, fearing popular reaction, never even declared war on the USSR. Soviet diplomatic representatives remained in Sofia through it all.

In August 1943, Tsar Boris died mysteriously a few weeks after returning from a final meeting with Hitler in Germany.[2] The massive crowd at his funeral, the largest yet assembled in Sofia, mourned not only his passing, but also the fate that had left them leaderless in a war that was now lost. Anglo-American bombings of Sofia and several other Bulgarian cities began in November 1943, further shaking morale on the home front. The Filov regime had foolishly declared war on both Great Britain and the United States in 1941. By 1944 Yugoslavia's massive Partisan movement under Tito was strong even in Macedonia. In Bulgaria proper, the small resistance movement organised by the Bulgarian Communist Party since 1942 was now staging guerrilla attacks. Soviet troops were fast approaching. Before the government could leave the war, units of the Red Army entered Bulgaria on 8 September, 1944. The Communist resistance seized power in Sofia the following day.

The political history of the German alliance was more complex than this summary of military events might suggest. Yet Boris's appointment of Bogdan Filov as Prime Minister in February 1940 placed a stubborn supporter of that alliance in a decisive position until the summer of 1944. Filov was an archaeologist by training and had served in no official positions other than rector of Sofia University and President of the prestigious Bulgarian Academy of Sciences. Basically a narrow nationalist and an undiscriminating admirer of all things German, Filov lacked any coherent vision for the fascist restructuring of Bulgarian society, the economy included.[3] This was undoubtedly one reason why Boris picked him to replace the belatedly independent Kioseivanov, rather than the genuinely fascist political leader Aleksandur Tsankov or the maverick general Khristo Lukov. The tsar and the army's general staff still wanted no mass fascist movement and no fundamental changes in an internal order based on 'non-party', authoritarian regimes since 1934.

Thus the Filov regime, though fatally binding the country and the dynasty to the German fortunes of war, operated under a more limited, ambiguous mandate on its own territory. By mid-1941, the occupation of Macedonia brought the regime to the height of its popularity. Despite setbacks that year, the Communist underground was able, by 1942, to organise a fledgling Fatherland Front of opposition groups. In 1943, opposition from a variety of sources, ranging from Communist sympathisers to Tsar Boris himself, was able to frustrate the plans of the Filov regime to put

Bulgaria's small and long-respected Jewish population of over 50,000 into the pipeline for evacuation to the Nazi death camps.[4]

General Lukov's assassination in February 1943 and Boris's own sudden death a few months later did not strengthen the regime's authority as Filov had expected. He became co-regent with the tsar's uncle Kiril and the pro-German General Mikhov in September 1943. Filov did succeed in naming his pliant Finance Minister, Bozhilov, as Prime Minister, but was unable to stem growing criticism in the Subranie or to crush the underground Fatherland Front. The war was simply going too badly. Communist morale grew with every Soviet victory. The Anglo-American bombings prompted some of Filov's associates to press for peace feelers to the Western allies. Filov hestitantly approved, but then drew back. Such inquiries were too late, when they finally came from a cabinet headed by the Agrarian Ivan Bagrianov. He had replaced Bozhilov in May 1944. Bagrianov's continued co-operation with German authorities and his refusal to deal with the Fatherland Front led to his replacement by Konstantin Muraviev and a pro-Western cabinet barely a week before the Red Army arrived. Bagrianov's economic initiatives to halt food shipments to Germany and to hold down inflation came too late to have much effect.

The economic policies pursued in wartime Bulgaria were therefore almost entirely those of the Filov regime, but were implemented without full political authority. Both internal opposition and German interference, especially in the occupied territories, played an inhibiting role as well. The regime's efforts to include industry and agriculture in military mobilisation aspired to central planning. They were ironically frustrated in important ways through Bulgaria's bilateral relationship with the German war economy. Peasant smallholders and small-scale manufacturers and traders also posed the regime problems which became worse as the war continued and the economy became more isolated.

Two principal economic issues emerged from these events. In what respects was the way prepared or not prepared for post-war central planning on the Soviet model? What sort of growth had this wartime economy been able to achieve?

State Supervision of Agriculture

For agriculture, the demand prompting initial growth came from the German war economy and from the efforts of Bulgarian state agencies to extract greater deliveries from peasant smallholders and the co-operative network. Yet further growth in 1943–4 followed from peasant efforts to avoid these two sources of non-market demand.

Sales to Nazi Germany, almost all agricultural except for some copper ore, rose to 72 per cent of the total Bulgarian export value for 1941–2 and to 83 per cent for 1943 and the first 8 months of 1944. In the process, the real value of Bulgarian exports for 1941–3 averaged 25 per cent more than the 1939 figures.[5] The structure of these sales continued generally along pre-war lines. Tobacco was the principal commodity, rising slightly to reach 42 per cent of export value in 1943. The price paid, however, was less than in 1939. Eggs, soya beans, fresh fruit and vegetables declined, from 29 per cent to 13 per cent combined. Processed fruit pulp recorded the one dramatic increase. It climbed to 11.1 per cent of export value by 1943, from 3.8 per cent in 1939.[6] Thus a principal export in the post-war era first became significant during the Second World War.

The German role in pulping and other agricultural modernisation was, however, minimal. Only the state monopoly on tobacco production, already taken over from French interests before the war began, afforded German authorities any direct control over crops. The Reemstma trade organisation managed one-third of total production through the Bulgarian co-operative network. But cigarette manufacture received no more attention than in the 1930s. The best post-war study of Nazi food management across Europe finds no other successful enterprises or programmes, with the exception of the same I.G. Farben project for promoting soya-bean cultivation that was at work in Romania.[7] Yet the output and yield of soya beans fell off sharply after their 1941 increase. The Sudostropa enterprise was unable to spread flax cultivation for its processing plants in the Dobrudja. German experts were lacking, and the weather was bad. The Buschag corporation failed in its cattle-breeding efforts, nor could it convince the Bulgarian Ministry of Agriculture to support its project to cross Bulgarian with Merino sheep. Only a few peasants were trained in the use of modern farm machinery at two model German villages.

Grain exports to wartime Germany had in any case never recovered from the disastrous Balkan drought of 1942. Even imports of German grain proved insufficient to prevent the start of bread rationing. Bulgaria's 1943–4 export of grain to Germany amounted to only one-half of the 1939–40 level. Had deliveries of German tractors and other farm machinery arrived in the quantities promised, however, the Bulgarian capacity for grain exports might not have continued its inter-war decline.

Grain was the centrepiece of state efforts to mobilise Bulgaria's economic resources for the war effort. And, as might be expected, the already existing grain-export organisation, Hranoiznos, was the most important and effective of the several responsible state agencies. Its monopsony powers in buying five crops for export or domestic sale in 1939 (cotton and hemp as well as wheat, rye and flax) grew to 23 by 1943. Tobacco and fruit were the only significant omissions. The pre-war practice of operating through local agents and co-operatives to buy whatever they offered for sale had evolved by the bad harvest of 1942 into *ad hoc* powers to requisition grain for the army and the urban population. Norms for delivery per hectare were set according to local conditions. The regime's Council of Ministers charged Hranoiznos with similar collections for other crops. A variety of other agencies were soon attempting to control meat and egg sales in this fashion.[8] It was, nevertheless, Hranoiznos upon which the regime's Central Agency for Special Requisitions relied in order to collect foodstuffs for supplying the German army, first for its 1941 attack on Yugoslavia and Greece, and later for its occupation troops.

The efforts of the Filov regime to set up several new agencies to co-ordinate agricultural production with the rest of the country's economy were less successful.[9] The War Ministry created a Directorate for Civilian Mobilisation (DGM) in May 1940, with powers to mobilise all economic enterprises for the war effort. All were required to adhere to yearly production plans, but individual enterprises were left to draw them up; less than half even submitted plans. Then the DGM joined the Ministry of Agriculture in preparing a more ambitious Five-Year Plan for 1942–6. It proposed measures to raise the cultivated area and to consolidate scattered parcels into unified units of cultivation. This plan for continuing mobilisation also foresaw complete state control of agricultural prices and income. But repeated requisitions of grain, in particular, beyond the annual quotas for the mobilisation plan, discouraged peasant co-operation.

The Directorate for Foreign Trade, established in June 1940, was meanwhile attempting to supply the agricultural sector with additional credit and badly needed, affordable fertiliser, as well as other imported inputs through a fund for price equalisation. The fund set aside a share of export earnings for subsidising cheaper imports. Any chances that the DGM might have had to bring these agencies under co-ordinated control disappeared in 1943 with the army's creation of a High Command for the War Economy. Its functions largely duplicated those of the DGM. The new agency reflected the general staff's impatience with any plans not geared to meeting the military's short-term requirements. These needs were growing in 1943. German pressures led the Bulgarian army to expand its role in occupied Yugoslavia and Greece.

The Wartime Agricultural Record

Wartime agricultural production could record real net growth over the 1939 level only in 1941. That year's modest increment of 3 per cent derived from a surge in animal numbers, plus egg and dairy production. According to Table 4.1, gross crop production did rise from a low point in 1940, the first year of mass army mobilisation, but came no closer to the 1939 level than 91 per cent in 1943.

More significant for the modernisation of Bulgarian agriculture was the further shift in crop structure along lines that had first become visible during the First World War. The share of cereal grains fell from 48.5 per cent to 35.9 per cent of real crop output between 1939 and 1944, despite the re-acquisition of the grain-growing southern Dobrudja in 1940. Industrial crops, primarily tobacco and cotton, saw their share drop slightly from 14 per cent to 12.4 per cent. Sharply reduced yields cancelled out 60–80 per cent increases in cultivated area. Fruit and vine output, up from 9.3 per cent to 11.9 per cent, drew virtually even with industrial crops. Table grapes were largely responsible for the surge in fruit production noted in Table 4.1, from 1942 forward. The even larger increment for vegetable production boosted its proportion of total crop production, from 18.3 per cent to 29.5 per cent. A tremendous upsurge in potato and kidney-bean cultivation accounted for over 80 per cent of the 1943 volume of vegetable production. Hectarage for beans, potatoes and sugar beet all showed sharp increases. Tomato and green-pepper cultivation, so prominent in Bulgaria's more recent exports, remained relatively small.[10]

How can we explain the wartime diversification of Bulgarian

Table 4.1: Indices of Agricultural Production,[a] 1939–44 (1939 = 100, in 1939 prices)

	Grain	Vegetables	Industrial crops	Vineyards, fruits	All crops	Net animal
1940	72	72	105	44	78	102
1941	69	107	104	64	84	129
1942	40	103	88	101	68	130
1943	73	140	84	106	91	63
1944	60	131	72	104	81	77

[a] Gross production minus seed deductions.

Source: J.R. Lampe and M.R. Jackson, *Balkan Economic History, 1550–1950* (Bloomington, Indiana: Indiana University Press, 1982), table 13.5, pp. 533–7.

crops that did occur? Conscious German encouragement does not furnish the answer. The fast-growing potato and bean production of 1941–3 was largely requisitioned for the Bulgarian army or sold to the urban population. Barely 5 per cent was exported to Germany. For fruit, its processing into pulp for easy transport did send a majority of the relatively small strawberry and apricot production to the Reich. But Nazi officials, especially those in the party's powerful Foreign Office, had discouraged such processing in Bulgaria until 1943, arguing instead that the Reich and especially Vienna should be its centre.[11] By the time they changed their minds, Bulgaria's processing capacity had already increased, as we shall see, on its own account.

More than anything else, the increase in bean and grape cultivation constituted a reaction by peasant smallholders to the exactions which the mobilisation agencies had placed on *other* sorts of agricultural production, especially grain. By the end of 1941, Hranoiznos found itself with insufficient grain reserves. Too much had been exported to Germany in 1940 and too much given to the Bulgarian army, now over 400,000 men, in 1941. A bad harvest following the drought of 1942 left the agency to choose between inflationary price increases and forced requisitions in order to meet military needs. Less than 90 per cent of requisition quotas were actually collected. Peasant smallholders reacted by holding back grain from the agency's threshing inspectors and German observers. Some of this grain could then be sold for prices so high on the black market that was growing in Bulgarian cities that the rest of the grain could be kept for consumption within the household.[12] Rising land prices were also discouraging expanded grain cultivation. Agricultural land continued to be sold, typically in order to cover outstanding debts, but it was becoming too expensive to use for any cultivation that was not intensive. Only the 58 producers' co-operatives operating by 1943–4, complete with consolidated fields and tractor fleets, were able to record high yields (twice the national average) and to earn a large, legal income from grain cultivation.[13]

Peasant smallholders had turned to industrial crops during 1941, and especially to milk production, as more profitable alternatives. By 1942–3, low official prices and high requisition quotas had reduced the attraction. A growing fodder shortage made livestock more expensive to maintain, in any case. The remaining alternative was fruit and vegetable production. Both commanded better requisition prices and were readily disposable on the country's urban markets, unlike tobacco and most other industrial crops.

This diversification may well have contributed to the growing shortage of bread and meat in Bulgarian towns. Bread rationing and meat restrictions had begun by 1942. Meatless days and daily bread rations, reduced to 300 grams, soon followed. The prices of these two commodities led the way among domestic items. The inflationary spiral pushed up official food prices in Sofia by 563 per cent over the 1939 level by 1944, and by 738 per cent on the black market.[14] Urban consumers, especially in the capital city, paid the largest part of the price for this further diversification of Bulgarian agricultural production.

Inflation and the German Alliance

The cost of Bulgaria's wartime alliance with Nazi Germany fell more heavily on industry and other sectors of the economy than on agriculture. This greater burden may be judged from rates of inflation which were significantly higher than for agriculture. The higher prices were the direct result of Bulgaria's participation in the German war effort. The Filov regime found itself saddled with high costs and little profit in the occupied areas of Macedonia and Thrace. Within the old borders, access to badly needed German imports proved too limited to prevent crippling shortages and hence inflationary pressures.

The reincorporation of the southern Dobrudja in 1940 and then the addition of Macedonia and Thrace in 1941 had increased

Bulgaria's population by 35 per cent. The total was 8.6 million, almost the present figure. Territory increased even more, by 49 per cent. Yet arable land in this new territory augmented total arable by only one third. Most of it was more suitable for tobacco or cotton than grain crops. For Thrace, German pressures obliged Bulgarian authorities to divert grain and coal supplies to support the local population as well. Little cotton was collected for Bulgarian markets in return. By 1943, the Filov regime had to use half of its available hard currency to import Turkish cotton.[15]

For Macedonia, the Bulgarian side supplied a variety of industrial outputs, but found the anticipated market for its own finished goods pre-empted by German manufactures, which were imported free of any Bulgarian duties. The terms of the April 1941 agreement with the head of the economic section in the German Foreign Ministry, Karl Claudius, spelled out some other unfavourable arrangements. In return for doubling the German payment to Bulgaria for supporting the former's troops, the secret agreement stipulated major concessions from Sofia. The Bulgarian side would finance German exploitation of Macedonia's valuable chrome and other mines, pay for German requisitions, and cover some outstanding Yugoslav and Greek bank debts in Macedonia and Thrace. It would also accept German army script and cover the cost of German military construction in Bulgaria proper.[16] The Macedonian mine concession alone yielded the German war effort annual deliveries that were 30 times Bulgaria's 1939 ore exports.

German economic relations with Bulgaria proper are more difficult to appraise. In order to do so, we must enter what has rightly been called the maze of agreements between the two countries. Their variety reflects the absence of a fully co-ordinated German programme for dealing with the Bulgarian economy. This was hardly surprising, given the early chaos of the Nazi war effort. The second German-Bulgarian agreement, signed in October 1940, was the first during wartime. The strong German bargaining position after the fall of France forced an important concession from the Bulgarian side. It agreed to accept payment for Bulgarian exporters in new notes emitted by the Bulgarian National Bank, if German imports were not immediately forthcoming to balance the transaction. This concession proved no problem until the last four months of 1941. Then the Bulgarian export surplus, which was to persist in German trade for the rest of the war, appeared for the first time.

The German response was to press for Bulgarian reductions in the import tariffs applied to German goods. This was their representatives' constant refrain in the nine meetings of the German-Bulgarian Trade Commission held between 1940 and 1944. At first, Bulgarian agreement to several tariff reductions cost the national economy only import duties foregone. Later Nazi demands for still more reductions helped the German side to evade its responsibility to deliver specifically promised goods, especially agricultural machinery and fertiliser, let alone enough to close the trade deficit with Bulgaria. By the fourth session, in October 1941, the Bulgarian side was refusing to credit the German clearing account with the 35 per cent premium on the former's earnings from exports to other countries which transited Germany. In retaliation, German representatives denied Bulgaria the right to transfer earnings from past exports to Germany into hard currency that could be spent elsewhere, without specific German approval. Further Bulgarian complaints at these meetings centred on undelivered military and agricultural equipment, the burden of missing fibre and oil imports on the Bulgarian industry, and the loss of tariff revenues. One Bulgarian survey in 1942 found that German imports had been delivered in only 28 of 132 agreed categories. Prices for the few arrivals were 50-60 per cent over those stipulated by the trade commission protocols.[17]

When Bulgaria's bargaining position for its increasingly valuable agricultural exports improved late in the war, the result was not a significant improvement in the terms of its essentially German trade. The slight improvement for 1943 reflected in Table 4.2 still left the ratio of export versus import prices at 61 per cent of their 1939 level. By 1944 the ratio had fallen to 56 per cent. Shortages of promised German imports combined with the diversion of 15-20 per cent of Bulgarian agricultural production to the domestic black market to exert greater upward pressure on all prices.

Pressure on the Filov regime to increase the emission of currency was even greater. German support payments for their own troops repaired what would have otherwise been a large deficit in the Bulgarian balance of payments. But no German loans were forthcoming, as at the start of the First World War, to help Bulgaria finance its own war effort. The Filov regime was forced to cover these expenses mainly by emissions of new bank notes from the National Bank. The Agricultural Bank lent large sums to the regime from 1941 onwards, but these proved insufficient. As a result, notes in circulation rose 629 per cent to cover the majority

Table 4.2: Foreign Trade and Inflation Indices, 1939–44 (1939 = 100)

	Constant 1939 prices		Current prices		Terms of trade	BNB note circulation	Retail prices[a]
	X	M	X	M			
1940	85	112	158	103	66	142	113
1941	85	123	239	124	54	276	151
1942	80	127	310	174	56	412	196
1943	82	125	352	215	61	629	252
1944	27	74	448	253	56	1,180	379

X = exports; M = imports.
[a] List of 34 items.

Sources: Lampe and Jackson, *Balkan Economic History, 1550–1950*, tables 13.3 and 13.8, pp. 526–9; N. Todorov et al. *Stopanska Istoriia na Bulgariia, 681–1981* (Sofia, 1981), p. 427.

of the fivefold increase in budget expenditures for 1939–43. By 1944, note emission had nearly doubled again. Gold coverage of the leva dropped from a pre-war 26 per cent to 4 per cent. The belated efforts of the Bagrianov government to compensate at least state employees for the rising cost of living, with 50–100 per cent salary increases, added one last impetus to the inflationary spiral.[18] This inflation was one major symptom of the strain that association with the German war effort imposed on the Bulgarian economy.

Industry and the German Alliance

The peculiar pattern of Bulgaria's industrial growth was the other major symptom of that strain. For manufacturing, real net output was 18 per cent ahead of 1939 by 1941, but fell back to the pre-war level by 1943. By then only food, wood, paper and electrical output were still ahead of the 1939 levels. Mining, essentially of coal and copper, was clearly responsible for the better overall growth of gross industrial output noted in Table 4.3. It rose steadily to 71 per cent more than its 1939 level by 1943. The overall increase of 10 per cent was not enough to prevent the industrial share of Bulgarian national income from slipping slightly to less than 8 per cent.

Setting back the process of modern industrial development was the further spread of small-scale enterprises, just meeting the definitional minimum of 10 workers and 10 horsepower. The

Table 4.3: Indices of Industrial Production, 1939–44 (1939 = 100, in 1939 prices)

	1940	1941	1942	1943	1944
Gross output	111	118	113	110	90
Investment goods	99	91	102	97	97
Food and tobacco	123	150	136	112	106
Textiles	97	92	67	66	58
Metals	81	99	112	89	77
Mining	119	126	157	171	184

Source: Lampe and Jackson, *Balkan Economic History, 1550–1950*, table 13.12, pp. 562–3.

number of enterprises rose by one-third, adding almost a thousand to the 1939 total of 3,100. Most were founded to take advantage of rising prices for manufactures and to escape DGM controls on larger firms. Their small size reduced the average number of workers for all firms from 29 in 1939 to 26 at the end of the war. Most were partnerships or single-owner enterprises, rather than joint-stock corporations with their greater potential for capital accumulation. A majority of the new firms were set up in Sofia, making the capital's share of all Bulgarian industrial enterprises more than 40 per cent by 1942.[19]

Mechanical horsepower per worker, on the other hand, grew enough to outweigh by a small margin the reduction in average firm size. For metal manufacture, the growth of horsepower was striking, from less than one to three per worker. Only textiles experienced a slight decline. To reconcile even this small increase in capital formation with declining firm size, we must refer again to the indirect effects of economic relations with Germany, rather than to the direct effect of the Bulgarian regime's programme of wartime mobilisation.

That programme required annual production plans from all designated enterprises by 1942. Barely 40 per cent of 1,400 industrial firms had even submitted plans. The lack of state credit and raw materials also plagued the industrial part of this first 'central plan'. Similar limitations frustrated specific programmes in 1942 and again in 1943 to establish new state enterprises for caustic soda, calcium and even synthetic rubber production. Then the start of Allied bombing prompted the removal of raw materials and finished production from Sofia factories, at least as much as the rail network, a primary bombing target, could still carry.

By 1944, the share of state enterprises in industrial production was just 5 per cent, compared with 9 per cent in 1939. Most of this production came from the several state coal mines. Here was the one branch of industry where employment had expanded more rapidly than output. Some 40 per cent of these 1,800 miners had been exempted from military service, well beyond the 9 per cent average for industry.[20] Otherwise, the state's primary role in Bulgarian industry remained that of a contractor. It drew on private firms to supply military provisions and to support a civilian bureaucracy that had grown by one-half during the war.

Those firms with state contracts tended to be larger and relatively better supplied with raw materials. Best off were the aforementioned plants for processing fruit pulp. They nearly doubled in numbers to reach 76 by 1944. Real output rose sixfold, labour and horsepower even more. Their 3,400 workers by 1944 represented almost 4 per cent of the total workforce in manufacturing.[21] The German purchasing agency for processed food bought the bulk of their output.

Other Bulgarian manufacturers, even those steadily receiving state contracts, had a different sort of experience with the German war effort. First there was the joint German-Bulgarian Industrial Commission of 1941–2. Its actual purpose was to *prevent* native industry throughout south-eastern Europe from producing anything that might compete with German manufacturing, and to control the rest through a single comprehensive cartel. Like their counterparts in Hungary, Bulgarian representatives postponed enough meetings and evaded enough commitments to make this scheme a dead letter by 1942.[22]

A more serious burden for Bulgarian industry was the German failure to deliver promised imports of raw materials and semi-finished inputs. The imported share of Bulgarian industrial inputs declined from 29 per cent in 1937–8 to 14 per cent for 1941–3 because of these shortfalls. By 1944, the real value of metal and machine production had fallen to 43 per cent of the 1939 level, in large measure because total inputs were just 24 per cent of what they had been. Chemical production had in similar fashion fallen to 75 per cent. Inputs were down by one-half. Despite cotton fibre imported from the Soviet Union in 1940–1 and from Turkey in 1943, textile production had dropped to 63 per cent of its 1939 level by 1944.[23]

Compounding these shortages was the diversion of 18,000 skilled workers, about one-fifth of the industrial labour force, to work in Germany itself.[24] Their absence, together with the Bulgarian army's mobilisation of over 400,000 troops, created a serious shortage of skilled labour. That shortage persisted into the early post-war period (see Chapter 5).

What the German side did supply to Bulgarian industry, at least for 1940–1, was a variety of manufacturing machinery and equipment, though typically not the latest technology, which totalled 20 per cent of import value. This was twice the pre-war fraction. By 1942–3, the fraction had fallen to 5 per cent. Military machinery and equipment now predominated. The earlier deliveries were hardly part of any Nazi plan to encourage Bulgarian industry. They were simply goods which were not essential to the initial German war effort and whose sale to Bulgaria would reduce the imbalance in their bilateral trade. Ironically, the German-encouraged shortage of skilled labour also helped to promote the introduction of such imported, often labour-saving equipment.

To this limited extent, the wartime Bulgarian alliance with Nazi Germany accidentally laid some groundwork for the industrialisation drive launched by the post-war Communist regime. On the other hand, the state's wartime controls and the further spread of smaller firms weakened the efficiency of the larger, more mechanised industrial enterprises in private hands. It was this weakness, rather than the strength of private industry, which both hindered nationalisation and made it a more rational policy under the new Communist regime.

Notes

1. These prophetic words are cited in both the major Western studies of wartime Bulgaria: Marshall Lee Miller, *Bulgaria during the Second World War* (Stanford, California: Stanford University Press, 1975), p. 55; and Hans-Joachim Hoppe, *Bulgarien: Hitlers einwilliger Verbundeter* (Stuttgart, 1979), p. 124. A recent and lengthy Bulgarian account of relations with Germany is Vitka Toshkova, *Bulgariia i Tretiiat Reich, 1941–1944* [Bulgaria and the Third Reich] (Sofia, 1975).

2. Boris had again refused to send Bulgarian troops to the Russian front. Bulgarian public opinion both then and now has little doubt that the Gestapo played some part in his death. See Miller, *Bulgaria*, pp. 133–48; and Ilcho Dimitrov, 'Smurtta na Tsar Boris III' [The Death of Tsar Boris III], in his *Bulgariia na Balkanite i v Evropa* (Sofia, 1983), pp. 124–57, for a convincing rejection of the flawed evidence for this view.

3. Miller, *Bulgaria*, pp. 89–92. Also see Filov's diary as translated in part by Frederick W. Chary in *Southeastern Europe*, vol. I, part 2 (1974) and vol. II, parts 1

and 2 (1975).
4. See Frederick W. Chary, *The Bulgarian Jews and the Final Solution, 1940–1944* (Pittsburgh, Pennsylvania: University of Pittsburgh Press, 1972). On the wartime experience of the Communist Party, see Nissan Oren, *Bulgarian Communism: The Road to Power, 1934–44* (New York: Columbia University Press, 1971).
5. John R. Lampe and Marvin R. Jackson, tables 13.2–13.3, *Balkan Economic History, 1550–1950: From Imperial Borderlands to Developing Nations* (Bloomington, Indiana: Indiana University Press, 1982), pp. 524–7.
6. *Spisanie i izvestiia na Glavna Direktsiia na Statistika* [Journal and Bulletin of the Chief Board of Statistics], vol. I (1945), pp. 29–45. Also see David Koen. 'Bulgarska industriia i imperialisticheskata Germaniia prez ftorata svetovna voina' [Bulgarian Industry and Imperialist Germany during the Second World War], in Khristo Khristov, V. Hadzinikolov, *et al., Bulgarskoto-Germanski otnosheniia i vruski* [Bulgarian-German Relations and Ties], vol. I (Sofia, 1972), pp. 526–9; Nikolai Todorov *et al., Stopanska istoriia na Bulgariia, 681–1981* [Economic History of Bulgaria, 681–1981] (Sofia, 1981), pp. 415–24.
7. See Karl Brandt, *Management of Agriculture and Food in the German-occupied and Other Areas of Fortress Europe* (Stanford, California: Stanford University Press, 1953), pp. 198–212.
8. S.D. Zagoroff, Jeno Vegh and Alexander D. Bilimovitch, *The Agrarian Economy of the Danubian Countries, 1933–1945* (Stanford California: Stanford University Press, 1955), pp. 387, 428–38.
9. Ibid., pp. 418–27; Todorov *et al., Stopanska istoriia na Bulgariia*, pp. 413–15.
10. Lampe and Jackson, *Balkan Economic History*, table 13.5, pp. 533–7; in Todorov *et al., Stopanska istoriia na Bulgariia*, table 40, pp. 458–9.
11. Dietrich Orlow, *The Nazis in the Balkans* (Pittsburgh, Pennsylvania: University of Pittsburgh Press, 1968), pp. 131–9. On the structure of Bulgarian agricultural exports to Germany, see Zagoroff, Vegh and Bilimovitch, *Agrarian Economy*, tables XIX–XX, p. 461.
12. The way in which peasants shook their sheaves into blankets, thereby extracting up to one-quarter of the grain for themselves, before taking the sheaves to official threshing stations is described in Irwin T. Sanders, *Balkan Village* (Lexington, Kentucky: University of Kentucky Press, 1949), pp. 186–8. On wartime agriculture in general, see Todorov *et al., Stopanska istoriia na Bulgariia*, pp. 413–28.
13. Grigor Popisakov, *Ikonomicheska otnosheniia mezhdu NR Bulgariia i SSSR* [Economic Relations between the PR Bulgaria and the USSR] (Sofia, 1968), pp. 257–66.
14. Lampe and Jackson, *Balkan Economic History*, table 13.5, pp. 533–7; Zagoroff, Vegh and Bilimovitch, *Agrarian Economy*, pp. 423–6, 433–7.
15. Hoppe, *Bulgarien*, pp. 132–3; Koen, 'Bulgarska industriia', p. 516.
16. Popisakov, *Ikonomicheska otnosheniia*, pp. 37–8.
17. Koen, 'Bulgarska industriia', pp. 499–516.
18. Todorov *et al., Stopanska istoriia na Bulgariia*, pp. 425–8; Miller, *Bulgaria*, p. 179.
19. Lampe and Jackson, *Balkan Economic History*, pp. 557–64; Liuben Berov, 'Kapitalobrazuvaneto v Bulgariia prez godinite na ftorata svetovna voina' [Capital Accumulation in Bulgaria during the Second World War], *Trudove na V.I.I. Karl Marx*, vol. III (1971), pp. 17–40.
20. Koen, 'Bulgarska industriia', pp. 510–13, 531–4.
21. Ibid., pp. 528–9.
22. Orlow, *Nazis in the Balkans*, pp. 134–4, 144–5.

23. Koen, 'Bulgarska industriia', pp. 514–43.
24. Hoppe, *Bulgarien*, p. 134.

5 COMMUNIST REVOLUTION, 1944–1947

Within hours of the Soviet declaration of war and the Red Army's entry into eastern Bulgaria, detachments of the Communist resistance movement had seized power in Sofia, in the name of the Fatherland Front. The date of this virtually bloodless coup, 9 September, 1944, has since become an anniversary whose celebration with a huge parade in Sofia symbolises the Communist accession to power. In fact, the consolidation of political power took several years. The economic revolution which accompanied and assured that consolidation took even longer. Hence this chapter's framework of four years to examine the interaction of Communist ideology with Bulgarian precedents for state initiative, existing government institutions and an immediate post-war economy which seemed in important ways to defy central control. Until the state's economic control was complete, at least for industry, central planning on the Soviet pattern could not really get under way.

The intial coup that toppled the week-old Muraviev regime attracted wide popular support beyond Communist ranks. The Red Army was given a genuinely warm welcome. A number of units in the Bulgarian army came over immediately to the Fatherland Front. With the replacement of only 800 officers and the addition of 700 Communist political officers, nearly half of an army of 450,000 men accompanied the Soviet divisions which pressed on quickly into Yugoslavia and Hungary. Bulgarian losses totalled 32,000 men killed, mainly in the bitter fight for Budapest in the winter of 1944/5. Simultaneously, the absent army ceased to be a factor in Bulgarian political life. So did those political leaders and groups associated with the wartime regime. Some 30,000 state employees were dismissed during September 1944. One-third soon faced trial and most of them received harsh sentences. Filov and his cabinet were tried and executed in 1945. The army's General Staff and a third of the pre-1944 officers were purged the following year. The young Tsar Simeon could attract few supporters. The vote to end the monarchy itself in September 1946 was overwhelming. Communist party membership had meanwhile jumped from 13,700 before 9 September, 1944 to 422,000 two years later.

The only potential contenders for political power besides the Communists were the three other parties in the Fatherland Front. The Agrarian, Zveno and Social Democratic representatives were never able to wrest the real leadership of either the Front or the immediate post-war government from Communist hands, despite holding three-quarters of the posts in the 1944–6 cabinet. They did subject its actions to public scrutiny and some criticism until 1947. Communist economic policy could not therefore proceed at a pace entirely of its own choosing.

These other parties faced insurmountable obstacles. Communists outnumbered all other party members combined on the local committees of the Fatherland Front. The Communist youth organisation was the largest by far. In November 1945, the renowned leader of the Bulgarian Communists, Georgi Dimitrov, returned from his 22-year exile in the Soviet Union to lend the weight of his international reputation to the Communist campaign for a single slate for the Fatherland Front in the elections eventually held in November. Such a slate appeared on the ballot and won 86 per cent of the actual vote, 73 per cent of the eligible electorate. Leaders of the other parties complained that the Communist-dominated People's Guard and People's Militia inhibited their campaigns for separate slates by threatening voters and disrupting meetings. Their 1945 appeals to the American members of the Allied Control Commission and to the US political representative in Sofia had won them the status of acceptable 'opposition parties', with permission to publish their own newspapers, and two postponements of the November elections. They could, however, obtain no further effective outside support from Americans after the postponements. The Anglo-American members were essentially observers on a commission controlled by Soviet officers and destined to be disbanded once a peace treaty was signed with Bulgaria in 1947.

Active American encouragement in 1945 only laid the opposition parties open to charges of collaboration with Western intelligence services. Recent Western scholarship casts doubt on these charges and makes clear the lack of Anglo-American military or economic leverage for influencing post-war Bulgarian politics. Both Western and Bulgarian research also suggests that the other parties in the Fatherland Front or in the emerging opposition suffered themselves from internal division and initial weakness.[1] The Communists suffered from neither of these.

There seems little doubt that from September 1944 and before, the Communist leadership was wholeheartedly committed to following the Soviet economic example of the 1930s. Its record of heavy industrial growth was after all unmatched elsewhere in Europe during the depression decade. On this basis some non-Communist support might be attracted. For Bulgarian Communists, such a system of central planning for nationalised industry and collectivised agriculture was as attractive ideologically to leaders who, like Traicho Kostov, spent the war years in Bulgaria, as it was to long-time exiles in the USSR, like Dimitrov, Vasil Kolarov or Vulko Chervenkov. Kostov, who was General Secretary of the party until Dimitrov's return, would have primary responsibility for economic policy during most of the immediate post-war period. He had no more desire to preserve private industrial or agricultural property than the others. Like them, he viewed the notion of opposition parties and a Western-style parliamentary system as an obstacle that would delay this economic transition. It would also deny the Communists the monopoly of political power to which any Leninist party has always aspired.

The presence of both the opposition parties and the Anglo-American members of the Allied Control Commission were admittedly obstacles which delayed the economic transition, if not much the political one. The latter was essentially complete with the appointment of Dimitrov as Prime Minister and of a cabinet at least half Communist in November 1946. A further and perhaps more formidable obstacle was the huge mass of peasant smallholders, the great majority of whom owned their land and owed their political allegiance to the reviving Agrarian movement. They had, moreover, emerged from the war free of the serious deterioration in living standards experienced by the urban population.

Against these obstacles, we must fairly note a number of longer-term advantages which the post-war context conferred on the Bulgarian Communist Party, advantages which Stamboliiski's Agrarians did not enjoy after the First World War. First, Bulgarian peasants and merchants could have no fond memories this time of the private pre-war market for agricultural exports, supported by access to international credit in return for adherence to the gold standard. The free-market mechanism, which was a chief attraction of European capitalist growth, had almost stopped working during the 1930s. Who could foresee its striking revival from the 1950s onward? Second, the Bulgarian economy and Bulgarian society received fewer permanent scars during the Second World War than during the First. The dead young men were far fewer, 32,000 compared with 155,000. No flood of several hundred thousand refugees poured in, requiring food, shelter and eventual absorption. This time, moreover, the loss of Macedonia seemed like an expected punishment for the Filov regime's alliance with Nazi Germany, rather than an injustice to be righted whatever the cost. As word of Nazi war crimes spread, some Bulgarians were probably drawn to accept the new Communist regime as a way of purging the past.

Finally, there was the Soviet Union. Bulgaria's population was traditionally pro-Russian. The Red Army undoubtedly offered the militia of Fatherland Front assistance against the Communists' political opponents in 1945–6. Beyond that, the USSR could promise Bulgaria the patronage which no Great Power had accorded it after 1918. Soviet political support would greatly reduce a reparations burden, which initially threatened to be as heavy as that of the Treaty of Neuilly. Soviet economic support also augured well. Already in late 1944, when Soviet policy was still primarily concerned with the Bulgarian war effort, the Red Army's mission in Sofia included an economic section with experts in agricultural and industrial planning. Imports of industrial raw materials like oil and cotton fibre had already proved useful in 1940–1. If agricultural and industrial machinery could also be provided, in return for Bulgaria's specialised agricultural produce, both sides might benefit. Only a Communist government in Sofia could hope to deliver this last advantage.

Nationalisation of Agricultural Trade

The collectivisation of Bulgarian agriculture went more slowly than any other part of the Communist economic programme for the revolutionary period 1944–7. By early 1949, only 11.3 per cent of arable land had been transformed from private peasant smallholdings into much larger collective farms, that is, producers' co-operatives (TKZS) on the pattern of the Soviet *kolhoz*. Most of the 1.1 million smallholdings surveyed in 1946 still survived. Decisive state leverage over the agricultural sector had none the less been established through the nationalisation of both internal and foreign trade.

Opposition to collectivisation came from several quarters and undoubtedly helped to slow its pace. The Red Army's economic advisers had tried to get the process off to a fast start in early 1945 by arranging the delivery of thousands of tractors from the USSR before the Second World War had ended. Yet one-third of the private farms and two-thirds of their arable land were holdings of 5 hectares or more, large enough to promise the household an adequate income.[2] Their peasant owners typically had no desire to give them up voluntarily. The opposition press of the Agrarian and Social Democratic parties emphasised this reluctance during the 1945–6 period of relatively free publication. The presence of the Anglo-American members of the Allied Control Commission may have further inhibited Communist initiative.

Some inhibitions also came from within the party itself. During the course of 1945, the Politburo discovered that it had overestimated the capacity of the party rank and file to promote a voluntary groundswell for collectivisation from the mass of smallholders. With party secretary Traicho Kostov taking personal control from the party's economic committee, the Politburo prepared a redistribution, in part to private smallholders, of all land in private properties over 20 hectares. Nearly half of this rather small amount of land, 3.6 per cent of total arable, was distributed to 129,000 smallholders in plots averaging 1 hectare by June 1946. None of the enlarged private holdings, however, was large enough to pose a potential threat to the Communist plan to use the existing framework of credit and other co-operatives to set up more TKZS than the 100 established in 1945. The other half of the redistributed land was contributed to these new farms. Liberal terms were set to attract private holdings. Not all of the holding need be contributed, and the collective would pay the former owner rent for land that was used. These two provisions, unique in post-war Eastern Europe, reflected the party's lingering hopes for a flood of voluntary members.

On this basis, however, only 3.8 per cent of arable land had been collectivised by 1947. Low delivery prices for what the collectives sold the state kept voluntary incentives to join low. Peasants began to slaughter livestock for lack of food by late 1946. It was in this troubled atmosphere that both Georgi Dimitrov and Kostov warned the rest of the party leadership in 1948, after the opposition press and the Allied Control Commission had ceased to exist, against 'illusions' about the imminent collectivisation of all agricultural land.[3] No one should have expected, Dimitrov went on, to create a socialist economy within the first five years.

The new regime had been able to take control of internal agricultural trade much more quickly. Serious food shortages and a widespread black market confronted the Communist-led Fatherland Front in all the major towns, especially Sofia, in September 1944. Heavy rains that fall and severe drought the next spring, followed by a lesser drought in the spring of 1946, made matters worse. Speculative hoarding spread. The net real value of crop and animal production for 1945/6 fell to 60 per cent of the 1939 level, down even from the 75 per cent recorded for 1943/4.[4]

Yet the Politburo's economic committee and the new governmental Higher Economic Council (VSS), set up under its auspices in early 1945, did have one great advantage. They did not need to create any new institutions to deal with this specific crisis or to increase their own authority in general. The Council simply assumed the powers of the wartime Directorate for Civilian Mobilisation (DGM), which it replaced. The DGM had continued to function after 9 September, 1944, as had the Hranoiznos organisation for controlling sales of 23 major crops (see Chapter 4). Under VSS direction, Hranoiznos used the wartime systems of delivery or production order, and of obligatory or forced deliveries to extract all it could. Local co-operatives rather than private traders were now charged with carrying out the delivery or production orders due from each property-owning peasant household. The Communist militia handled the outright requisitions. These were forms of 'military communism', according to recent Bulgarian scholarship, which recognises how much the peasantry resented these requisitions.[5] Rural resistance to collective farms surely hardened in the process.

Tobacco, fruit and some vegetables were the principal crops outside the wide purview of Hranoiznos. Urban shortages of fruit and vegetables in the spring of 1945 prompted the VSS, under party direction, to nationalise pulp processing plants. They had grown rapidly during the war in order to supply the Bulgarian army and the German home front. The network of rural co-operatives, which included some 1.6 million members, was instructed to requisition vegetables as well. The pre-war state organisation for vegetable and fruit export, Bulgarplod, was re-created and brought into the purchasing process. The Agricultural Bank and the co-operative network bought up 88 per cent of the 1945 tobacco crop for the state at fixed prices. By 1946, over 90 per cent of Bulgarian export value had been collected by state or co-operative organisations. These agricultural controls led the way in reducing the private share of internal trade to 57 per cent by 1947 and to 13 per cent by 1949.[6]

It remained only for a variety of state organisations to take over the role of the pre-war co-operative network after 1947. The co-operatives had served to manage almost half of this huge system for purchase and requisition in 1945−6. Without the prior extistence of this network and of Hranoiznos, however, it is doubtful that Communist authorities could have assumed control of agricultural trade with such relative ease. For foreign trade, no explicit decree for nationalisation was even necessary. It is also doubtful whether, without Hranoiznos and the co-operatives, gross real crop production marketed could have risen to 99 per cent of the 1939 level by 1948, up from 60 per cent in 1945.[7] At the same time, the 8 per cent increase in animal production over 1939 derived from the survival of private smallholdings more than from any state programme.

Economic Relations with the USSR

The new Communist regime faced another powerful pressure for agricultural recovery, beyond the food storage in the towns. This was the need to find exports to compensate the Soviet Union for the essential imports which the large, but war-ravaged Soviet economy was furnishing Bulgaria. Soviet political motives surely played their part. The attraction of assisting the Bulgarian Communists was obviously greater now that they were in power than in 1940−1, the only previous period of significant Soviet-Bulgarian trade. But now the Soviet need for compensation was also greater.

When that need had been less, the Soviet side had purchased more from Bulgaria than it had provided. The Soviet oil, cotton, metals and machinery furnished in 1940 amounted to 2.9 per cent of Bulgarian import value, compared with an export value of 4.6 per cent. Tobacco, pork and other agricultural goods made up the Bulgarian deliveries under the bilateral agreement signed in May 1940.[8] Trade in these goods resumed once the first post-war agreement was signed, soon after the Soviet Union accepted a formal truce with Bulgaria on 23 October, 1944. Its virtual barter terms specified the goods which both sides were to deliver. Subsequent annual agreements with the USSR in 1945 and 1946 have been followed by longer-term pacts since 1947.

The value of imports from the USSR accounted for 80 per cent of the reduced Bulgarian total for 1945–6. The commodities were crucial to the small economy's survival. The Red Army's command in Bucharest had already diverted some Romanian oil to Bulgarian use in 1944. Direct deliveries from the USSR began in 1945 and relieved a serious shortage in a country which had no oil of its own. Fibre from Soviet Central Asia allowed cotton textile production to recover its pre-war level. Thus ended the interwar Bulgarian effort (see Chapters 2 and 3) to use domestic cotton, except, ironically, for its promotion by Soviet advisers in the early 1950s (see Chapter 6). Soviet grain relieved the shortfalls created by the bad Bulgarian harvests of 1945 and 1946. Imports of fertiliser and agricultural machinery, especially tractors, assisted recovery and also facilitated the creation of the machine-tractor stations, which were then essential in Soviet-style collective farming. Most important among the machinery and metal products that made up the rest of Soviet imports were railway locomotives and rolling stock and an electric power station. This station, together with some equipment from Czechoslovakia, allowed the distribution of electric power to extend beyond Sofia again, for the first time since early 1944.[9]

Bulgarian exports to the Soviet Union were fully 95 per cent of total value in 1945. They declined to 61 per cent in 1946, while the imported percentage rose. From 1947 to the present, both have remained at just over 50 per cent. As may be seen from Table 5.2, these export fractions did not match imports from the USSR again until 1950. Bulgaria's overall trade balance was significantly positive only for 1945, and then only because of greatly reduced imports. The surpluses with other countries thereafter could not be converted into payment for the USSR. Hence, there was continuing pressure on the Bulgarian side to export even more to the Soviet Union.

Tobacco was by far the largest item; 64 per cent of sales went to the Soviet Union in 1945–6 and 80 per cent in 1947. The Bulgarian Agricultural Bank (BZB) easily arranged the sales through cooperatives, which were by then controlling tobacco leaf production. Some stocks of tobacco unsold in 1943–4 were also used. Greater diversification was clearly needed in the future. Fruit pulp and wine, however, were the only other commodities sold in significant amounts in these first post-war years, principally through Hranoiznos.[10]

Commerce with countries other than the Soviet Union, let alone

Table 5.1: Foreign Trade Volume and Terms of Trade, 1944–50

	Current prices (million pre-war leva) X	M	X/M ratio	Index (1939=100, in 1939 prices) X	M	Terms of trade
1944	11,357	6,478	1.75	74	27	56
1945	12,397	5,820	2.13			
1946	14,942	17,514	0.85			
1947	24,533	21,416	1.15			
1948	36,351	37,741	0.96	66	87	110
1949	369[a]	310[a]	1.19	74	118	
1950	797[a]	910[a]	0.88	102	102	71

X = Exports; M = Imports.
[a]In post-war leva.
Source: J.R. Lampe and M.R. Jackson, *Balkan Economic History, 1550–1950* (Bloomington, Indiana: Indiana University Press, 1982), table 13.3, pp. 526–9.

Table 5.2: Direction of Foreign Trade, 1938–50 (percentages)

	USSR X	M	Czechoslovakia X	M	Eastern Europe X	M	Other X	M
1938			5	6	13	18	88	82
1945	95	80			2	9	3	12
1946	61	82	11	6	17	9	17	9
1947	52	61	19	16	34	27	14	13
1948	52	58	11	12	29	25	19	17
1950	55	50	15	16	37	37	8	14

X = Exports; M = Imports.
Source: Lampe and Jackson, *Balkan Economic History, 1550–1950*, table 13.2, pp. 524–5.

genuine multilateral trade, got off to a much slower start in the immediate post-war period. The Bulgarian government signed bilateral trade agreements with Romania, Czechoslovakia and also Switzerland as early as 1945. The Czech connection was the largest and also the most valuable, because of access to more sophisticated machinery than the USSR could provide. There was little past historical or present political basis for trade with the United States, whose industrial economy emerged from the war undamaged as the world's strongest.

The one realistic opportunity for multilateral trade was the limited Balkan customs union proposed by Georgi Dimitrov in January 1948, after his warm final meeting with President Tito in Yugoslavia. Whatever the political rationale and problems of this proposal, its economic advantages would have been long-run rather than immediate. The potential members did not have enough industrial raw materials or manufacturing capacity between them to provide for all. Still, Romanian iron ore and oil and Yugoslav iron ore and non-ferrous metals would have helped Bulgarian industry by the 1950s. The Soviet rejection of Dimitrov's ideas as a 'cooked-up federation' came only five days after it had appeared in *Pravda*.[11] Stalin's split with Tito a few months later ended even normal Bulgarian trade with Yugoslavia by the end of 1948.

Consolidating Financial Control

Most of the financial sector had already been nationalised before the Communist regime came to power. Previous chapters have described the rise of the Agricultural Bank before and after the First World War, the resurgence of the central bank during the 1930s, and the merger of private provincial banks into the Banka Bulgarski Kredit in 1934. Their leverage had only increased during the war years. By 1944, their combined capital was 80 per cent of the Bulgarian total.

European-dominated commercial banks in Sofia comprised most of the remaining private sector. The German or Italian affiliates, including several banks simply taken over from their French or Belgian owners once the war had begun, became Soviet property under the terms of the Allied agreement at Potsdam in July 1945. The Soviet Union received 90 per cent of their stock, and they quickly ceased to exist. The one large Western European institution that tried briefly to establish its pre-war position was the famous French investment bank, Paris Bas. By October 1946, however, the bank's management had agreed to sell to the Bulgarian government its share in the Banque Franco-Belge, the United Tobacco Factory, and several textile firms. The original Communist plan to use the Banka Bulgarski Kredit to absorb private banks had not materialised; perhaps it proved unnecessary. When the 32 remaining private banks were merged with the central bank in December 1947, they held no more than 11 per cent of total bank capital and 6 per cent of deposits.[12]

The Bulgarska Narodna Banka was on its way to becoming the one important Bulgarian bank of the Communist era. The central bank's later absorption in 1951 of the co-operative and popular banks was to leave the previously powerful Agricultural Bank to languish. Already in 1948, the first in a series of state banks set up to invest in nationalised industry had failed to survive for more than a short period. The VSS and the Narodna Banka soon took over its activities.

Private financial assets outside commercial banking never recaptured their pre-war position. Western loans to the Bulgarian government were out of the question, and not just for political reasons. Bulgarian payment on its loans before and after the First World War had been suspended in 1940. French representatives were not able to arrange even partial payment until 1947. A broader agreement with all Western bondholders was negotiated in 1948, but never honoured.[13] No reparations were owed to the Western allies, according to the peace treaty signed in Paris in July 1947. Yugoslavia's Communist government forgave Bulgaria the 550 million leva ($25 million) owed for war damages, and Soviet pressure forced the Greek government to cut a claim twice as large down to $50 million. Its payment was left to fruitless bilateral negotiations. Thus the Western leverage over the Bulgarian state budget and the central bank that followed from the reparations issue after the First World War never materialised after the Second. Nevertheless the spectre of its possible reappearance served to encourage reliance on Soviet support.

The Communist-dominated government was meanwhile free to use both monetary and fiscal policy to soak up domestic funds still privately held. Large deficits in the state budget for 1945 and 1946 led the Narodna Banka to restrict new note emissions severely. Inflation was thereby better restrained than in neighbouring Greece or Romania. The cost of living did not even double from 1944 to 1947; it merely increased by 86 per cent, as note emissions rose only 13 per cent.[14] Credit in an economy still largely in private hands was also restricted. The entirely internal Freedom Loan for the 1945 state budget attracted some of the remaining private funds. Many more were appropriated in the monetary reform of 1947. Its provision blocked all private accounts, including those in state banks, over 20,000 leva, and imposed a one-time tax on the remainder. Its effect was to cut the Bulgarian money supply by two-thirds at a single stroke. Progressive taxes on individual income and excess profits soaked up potential new deposits. Their collection increased from 8 per cent to 27 per cent of state budget revenues between 1944 and 1947.[15] By then the Communist government had established virtually complete control of Bulgaria's financial resources. This proved to be a source of political as well as economic strength.

Problems with Private Industry

Private industry posed more problems for the Communist programme to consolidate economic power than did the financial sector. This was not surprising, given the nature of the average Bulgarian industrial enterprise. The war years had accelerated the inter-war tendency towards an increasing number of smaller and smaller firms, as noted in Chapter 4. A majority were not tied to the war effort or tainted by charges of speculation. Thus they were not vulnerable to nationalisation immediately after the war. But neither were they a likely basis for the modern, large-scale industry that the Communist leadership was committed to creating in the long run.[16] In the short run, Traicho Kostov and the party's other economic authorities concentrated on using all of the existing industrial capacity and restricting speculation or excess profits.

A majority of the industrial labour demobilised from the army or attracted from agricultural employment in the first post-war years found its way into the smaller factories and even unmechanised artisan shops. Total employment in industrial enterprises with more than 20 workers and 20 horsepower rose from 111,000 to 145,000 for 1944–7, but artisan employment climbed faster and nearly caught up, increasing from 80,000 to 127,000. The average size of firms meeting the minimal official definition of an industrial enterprise (10 workers and 10 horsepower) had already fallen to 24 by 1946. That year the Ministry for Industry needed 16 meetings to process the applications for new enterprises. The total number of enterprises had risen one-third by mid-1947 to pass 5,500. For 1944–7, the greatest changes were the declining size of the urban enterprise and the growing number of village shops. The average size of the former slipped from 59 to 49, led by Sofia which dropped to just 42. The number of artisan shops rose by one-half, because of growing numbers in villages and also in the Sofia area. Unmechanised lumber and brickyards might employ well over 20 workers, whereas the mushrooming metal shops typically employed three or less. By 1947, firms meeting even the official definition of an industrial enterprise had fallen from over 10 per cent to 3 per cent of the total number of metallurgical enterprises and shops.[17]

The immediate consequences of this deconcentration spelled trouble for the Communist Party programme. Agrarian opposition to the Communist rural policy of obligatory or actually forced food deliveries were now joined by complaints from the Union of Bulgarian Industrialists. They represented private owners who were loyal to the new regime, but found it increasingly difficult to operate efficiently. The VSS had raised authorised profits from 10 per cent to 15 per cent of sales volume in 1945, but actual earnings were usually less. Shortages of skilled labour and raw materials were responsible. The spread of smaller firms made both shortages much worse. Skilled metal or textile workers demobilised from the army or back from Germany were assigned to large enterprises, but they left these enterprises in droves to found their own small shops. National reserves of metals, cotton, rubber and other inputs amounted to one or two months worth in 1945. These shortages affected even food-processing plants. In Plovdiv there was the special problem of tobacco supplies, which had been moved out of town in 1944 to avoid Allied bombing raids. Now they could not be moved back, because of the lack of rail transport.

The ample supply of unskilled labour available once the army had been demobilised in 1945 could not therefore be fully employed. Industrial unemployment rose to 38,000 by 1946, or over 20 per cent of the workforce.[18] More than half of the unemployed were tobacco workers, many of whom were Communist Party members or sympathisers. If the party programme had promised anything to its followers during the 1930s, it was to eliminate unemployment. The appearance of that most lamentable capitalist phenomenon during the immediate post-war period must have made the transition to socialist industry on the Soviet pattern seem all the more urgent to the Bulgarian party leadership.

State Industry and Nationalisation

The industrial enterprises already under state ownership in 1944 and those nationalised in 1945–6 also encouraged this transition. They did so more because of the problems they faced than because of the successes they recorded. The size of the state sector, it may be recalled from Chapter 4, had fallen during the war years from 8–9 per cent to 5–6 per cent of industrial production. The mining of hard coal and the generation of electric power remained largely in state hands. State mines and power stations accounted for 89 per cent and 65 per cent of 1945 output respectively.

By 1947 the state industrial sector had grown in size to a prominent, though not commanding position. Its enterprises now included over half of total capacity, but only one-quarter of actual production. The conversion of the Varna shipyards from a German into a Soviet-Bulgarian joint enterprise and of the several arms factories, including the Lovech airplane works, into agricultural machinery plants, involved enterprises already under state ownership. The aforementioned acquisition of enterprises under Western European, German and Italian bank ownership served to nationalise tobacco-processing and the large Granitoid cement plant in Sofia, but little else. The largest addition to the state sector came from government offers to buy out firms producing necessities and from the confiscations decreed by the People's Courts.[19] They had been set up to deal with wartime collaborators. These tribunals were not to be trifled with; they handed down a minimum of 2,000 death sentences. There was no appeal to their decrees of confiscation for 137 industrial enterprises during 1944–5. By 1945 they were also empowered to take over any enterprises which defied the VSS decree banning any unexcused cutback in production.

The net effect of these confiscations none the less was minimal in a number of industrial branches. The seven textile firms taken over during 1945, for instance, accounted for less than 9 per cent of the joint-stock capital in the branch. More state leverage over textile production derived from the assigning of up to 8,000 workers, many of them unemployed, to factories designated in 1945–6 to process renewed imports of Soviet cotton. This method of production for state use with state-furnished raw materials closely paralleled the *ishleme* system of manufacturing cloth and shoes for the Ottoman army in the early nineteenth century.[20] No large new state enterprises were established during this difficult period.

For workers already employed in state industry, moreover, the experience of the immediate post-war period was not a happy one. Wages were low, and consumer goods extremely scarce. An urban food shortage, following the bad harvests of 1945 and 1946, reached its peak in 1947. Workers in state factories, and especially the hard coal mines, 'voted with their feet' during the summer months. They simply left their jobs to help cultivate the family smallholding and thereby assure their own food supply. Communist officials complained that 'we cannot hold winter workers once spring comes', because 'it is impossible to keep workers from crop cultivation'.[21] Fully two-thirds of the labour force left one coal mine. Several hundred army troops had to be sent in to keep another mine operating. Railway operation was curtailed because of the resulting coal shortage. During the winter months, some state enterprises were plagued with so many unexcused absences that regular hours of operation could not be maintained. Turnover was rapid. Sporadic strikes broke out. Even unskilled workers began to leave state plants manufacturing agricultural and other machinery and to seek out the small private metal works that were springing up. By 1947, the Communist regime was therefore facing a long-term threat to the coal and metal production that would be crucial to its vision of a modern industrial economy. It was also confronted with the last short-term problem which it had expected to face, a crisis in labour discipline.

Thus the Cominform decision to proceed more quickly with nationalisation of industry, taken at the meeting near Warsaw in September 1947 between Soviet and other Eastern Europe Communist leaders, was a welcome one for the Bulgarian party. None of its leaders could conceive of another solution to the problem they were facing in 1947. Nationalisation of industry on the Soviet pattern had always been their long-term goal anyway. Party Chairman Georgi Dimitrov called it 'the most important revolutionary step for the economy'. Now it was undertaken with all the urgency and secrecy of a military operation. By December of that year, 1,200 trusted party cadre had been assembled in Sofia and given several days of special training, all without public notice. Then came sealed instructions, and the party cadre positioned themselves around the country. At 11 o'clock on the morning of 23 December, 1947, they and local party leaders began entering the private industrial enterprises (numbering about 6,000) to announce the firms' nationalisation.[22] Public notices were then posted, and new Communist directors appeared the next day. There were no violent incidents or significant opposition. Over 90 per cent of the enterprises had less than 50 employees, and their average size was 23. These typically small firms could no longer act as a magnet to attract workers from the large state enterprises.

The problem now facing the Communist leadership was how to combine these small firms into the modern factories that they believed would assure the rapid development of the Bulgarian economy and solidify their political power. This they could not begin to do without the transition to Soviet-type central planning. That transition would last for at least a decade and is the subject of the next chapter. The tentative plan for industry and infrastructure in 1947–8 can then be considered in proper perspective, as a prelude to the three longer and increasingly comprehensive plans for 1949–60.

Notes

1. The Agrarians were badly split just as before the war. The Gichev faction was excluded from the front, compromised by the leader's participation in the Muraviev regime during the last week of the war. The *émigré* Agrarians were led by the other Georgi M. Dimitrov, not to be confused with the Communist leader of the same name. When he returned to Bulgaria in 1944 to attempt to unify the Agrarians once again, this Dimitrov could not overcome the handicap of his wartime association with British authorities in Cairo. Communist objections forced his resignation. A large group of left-wing Agrarians, led by Nikola Petkov, had originally opted to work within the Fatherland Front in 1944. Petkov and a part of the rapidly growing Agrarian membership total broke away the following year, charging Communist domination. The faction's membership had risen by 1946 to at least 50,000 and perhaps past 150,000; the figures remain in dispute. More clearly, Petkov's faction became the focus for opposition to Communist rule. Petkov organised a single slate of opposition candidates, which won 29 per cent of the vote in the October 1946 elections for a constituent assembly. The Communists, however, won 54 per cent and their Fatherland Front 70 per cent. Petkov's trial and execution in 1947, on questionable charges of plotting to overthrow the regime, symbolised the ruthless Communist determination to brook no public opposition or Western pressure once the Allied Control Commission had been disbanded.

The Zveno faction in the Fatherland Front sought to avoid these Agrarian problems by working behind the scenes. As in the mid-1930s, however, lack of mass support made their cabinet members, even the Prime Minister Kimon Georgiev, vulnerable to easy removal. The Social Democrats had been significantly weaker than the Communists since the 1920s. Like Zveno, their candidates had won less than 2 per cent of the vote in the 1946 election. The skilled industrial labourers who were their major support had been taken to Germany for war work, if they had not been drafted into the army. The agreement of a new, more pliant leadership to merge with the Communists in mid-1948 marked the end of organised political opposition in Bulgaria. A separate, but small and allied Agrarian Party has been allowed to remain to this day. The Bulgarian Communist Party (BKP) formally assumed that title again in December 1948, ending the long use of 'Bulgarian Workers' Party', adopted in 1927 during its own period of illegal opposition.

A balanced Western account of post-war political developments is Phyllis Auty, 'Bulgaria', in R.R. Betts (ed.), *Central and South East Europe, 1945–48* (Westport, Connecticut: Greenwood Press, 1971, reprint from 1951), pp. 25–51. On the British and American roles, see Elizabeth Barker, *British Policy in Southeast Europe in the Second World War* (London: Macmillan, 1976); and Michael M. Boll, *The Cold War in the Balkans: American Foreign Policy and the Emergence of Communist Bulgaria, 1943–1947* (Lexington, Kentucky: Kentucky University Press, 1984). The most accessible Bulgarian account is Mito Ususov, 'Formation of the Political

Structure of the People's Democracy in Bulgaria', in Isusov (ed.), *Problems of the Transition from Capitalism to Socialism* (Sofia, 1975), pp. 43–51. On the opposition parties, see Isusov, *Politicheskite partii v Bulgariia, 1944–1948* (Sofia, 1978). On the Communist Party, see Isusov, *Komunisticheskata partiia i revolutsioniiat protses v Bulgariia, 1944–1948* (Sofia, 1983); and Ilcho Dimitrov, 'BKP', in *Bulgariia na Balkanitei i v Evropa* (Sofia, 1983), pp. 307–28.

2. See S.D. Zagoroff, Jeno Vegh and Alexander D. Bilimovich, *The Agrarian Economy of the Danubian Countries, 1935–1945* (Stanford, California: Stanford University Press, tables 9–10), pp. 381–2. The change from 1934 was a slight decline for both farms and land over 5 hectares (see chapter 3). On the Soviet role in supplying and even drafting the decrees for the first labour collectives, see Angel Nikov, *Bulgaro-suvetski otnosheniia, 1944–1948* [Bulgarian-Soviet Relations, 1944–1948] (Sofia, 1978).

3. Nikolai Todorov *et al.*, *Stopanska istoriia na Bulgariia, 681–1981* [Economic History of Bulgaria, 681–1981] (Sofia, 1981), pp. 442–9; J.F. Brown, *Bulgaria under Communist Rule* (New York: Praeger, 1970), pp. 198–201.

4. P. Kiranov, 'Natsionalen dokhod na Bulgariia, 1939, 1944, 1945' [The National Income of Bulgaria, 1939, 1944, 1945], *Izvanredno izdanie na spisanie na Glavnata Direktsiia na Statistika* (Sofia, 1946), p. 85.

5. Mito Isusov, *Rabotnicheskata klasa v Bulgariia, 1944–47* [The Working Class in Bulgaria, 1944–1947] (Sofia, 1971), p. 45–9; Nikola Popov (ed.), *Ikonomika na Bulgariia* [The Economy of Bulgaria], vol. II (Sofia, 1972), pp. 78–9.

6. Todorov *et al.*, *Stopanska istoriia na Bulgariia*, pp. 434–42, 479; Popov (ed.), *Ikonomika na Bulgariia*, vol. II, pp. 103–6.

7. John R. Lampe and Marvin R. Jackson, *Balkan Economic History, 1550–1950: From Imperial Borderlands to Developing Nations* (Bloomington, Indiana: Indiana University Press, 1982), table 13.6, pp. 538–40. On the state absorption of private trade, see Zl. Zlatev, 'Sotsialno-ikonomicheska preobrazovaniia v Bulgariia, 1944–1948' [The Socio-economic Transition in Bulgaria, 1944–1948], *Istoricheski pregled*, vol. XXXVII, no. 1 (1981), pp. 7–19.

8. On 1940–1 trade, see Grigor Popisakov, *Ikonomicheska otnosheniia mezhdu NR Bulgariia i SSSR* [Economic Relations between the PR Bulgaria and the USSR] (Sofia, 1968), pp. 27–40.

9. Ibid., pp. 112–14; B. Simov, B. Blagoev and O. Oslianin, *Vustanovanie i razvitie na promishlenostta v NR Bulgariia, 1944–1948* [The Growth and Development of Industry in the PR Bulgaria, 1944–1948] (Sofia, 1968), pp. 117, 161–3.

10. Lampe and Jackson, *Balkan Economic History*, p. 544; Todorov *et al. Stopanska istoriia na Bulgariia*, pp. 474–8.

11. *Pravda*, 23 and 28 January, 1948.

12. Totiu Totev, 'Bankovata sistema na NR Bulgariia sled sotsialisticheskata revoliutsiia' [The Banking System of the PR Bulgaria after the Socialist Revolution], *Finansi i kredit*, vol. XXXI, no. 7 (1981), pp. 45–52.

13. It stipulated payment of interest arrears for 1940–8 and of 10–13 per cent of further interest from 1949 forward. William H. Wynne, *State Insolvency and Foreign Bondholders*, vol. II (New Haven, Connecticut: Yale University Press, 1951), p. 574.

14. Lampe and Jackson, *Balkan Economic History*, table 13.8, p. 552.

15. Todorov *et al.*, *Stopanska istoriia na Bulgariia*, pp. 480–5.

16. In the words of a recent Bulgarian study of the immediate post-war period, 'The small-scale and insufficiently profitable private economy was not in a position to secure large capital investments for the development of industry' (Zl. Zlatev, 'Socialist Industrialization in Bulgaria', in Isusov (ed.), *Problems of the Transition*, p. 176).

17. Lampe and Jackson, *Balkan Economic History*, Table 13.13, p. 564; Isusov,

Rabotnicheskata klasa, pp. 34–41, 87–8, 103–9. Also see Vishiia Stopanska Kamera, *Industrialni problemi v Bulgariia prez 1947 g.* [Industrial Problems in Bulgaria for 1947] (Sofia, 1947).

18. Isusov, *Rabotnicheskata klasa*, pp. 92–100, 186–96.
19. Simov, Blagoev, Oslianin, *Vustanovanie*, pp. 90, 168–78; Popov *et al.*, *Ikonomika na Bulgariia*, vol. II, pp. 74–5, 96–9.
20. Isusov, *Rabotnicheskata klasa*, pp. 25–30, 91–2. On the *ishleme* system, see Chapter 1.
21. Ibid., pp. 26–36, 92–8; Simov, Blagoev, Oslianin, *Vustanovanie*, pp. 106–10.
22. Petko Petkov, 'Restriction and Abolition of Capitalist Ownership in Bulgaria', in Isusov (ed.), *Problems of the Transition*, pp. 162–71.

6 THE FIRST FIVE-YEAR PLANS

The Bulgarian transition to a socialist economy on the Soviet pattern took longer than the immediate post-war period. Simply nationalising private industrial enterprises was not enough. A system of long-term central planning to co-ordinate outputs with inputs also had to be set in place. And for such planning to include all production, the collectivisation of small-scale, private agriculture seemed necessary. Bulgarian economists typically identify the date of its completion, 1958, as the end of the transition to the Soviet system. Yet the Third Five-Year Plan, which began that year, seems to have been as much a part of the Bulgarian effort to solve planning problems simply by adopting more of the Soviet system, as were the trial run in 1947–8 and the first two plans in 1949–52 and 1953–7. Like the first plan, the third was to be completed in less than five years. Misnamed by Western economists as a 'Chinese-style Great Leap Forward', the third plan (1958–60) drew primarily on Soviet practice and priorities.

Each of the first three plans followed a strategy of extensive growth. Huge amounts of capital and labour were funnelled into a few branches of industrial production. The five full-term plans since then have pursued more intensive growth, conserving labour, if not capital, as a way of increasing productivity. These plans still honour Soviet ideology, but have used a growing reliance on foreign trade as the basis for repeated reforms (see Chapters 7–9).

The present chapter can be brief for two reasons. First, the basic Soviet system for central planning is well known. So is Soviet development strategy: rapid growth of heavy industry to be achieved through concentrated investment from the state budget and a labour force augmented by peasant influx.[1] A smaller rural labour force is left on the mechanised collective farms to produce the surplus needed to feed a growing urban population. Bulgaria's post-1948 transition to this planning system and strategy followed the Soviet pattern perhaps more closely than did any of the other Communist governments in Eastern Europe. Second, although its general pattern is familiar, the period of the first three Five-Year Plans is the most neglected in modern Bulgarian economic history.

Western and Bulgarian economists have concentrated their efforts on the period since 1960, where reliable statistical evidence is more available and connections to the international economy more important. (Bulgaria published no statistical yearbook in the 1950s until 1956.) Western historians have typically not gone beyond the Second World War, nor have their Bulgarian or Soviet colleagues gone beyond the immediate post-war period.

Bulgarian political history from 1949 to 1960 also makes a detailed appraisal of these years more difficult. Accompanying several changes in Bulgarian party leadership was the Soviet transition from the Stalin to the Khrushchev eras. The single line of authority from party leadership to economic policy that is a hallmark of Soviet-style economies was doubtless present, but harder for outside observers to discern. Subsequent Bulgarian scholarship has trodden too lightly on these political links to make clear the inner dynamics of economic policy.

The period began with the illness and death of the party's respected leader, Georgi Dimitrov. A sick man at least from 1947 onward, he died in April 1949 after several months of treatment in the Soviet Union. Dimitrov enjoyed international prestige on the left as the eloquent defendant in the Reichstag fire trial, staged unsuccessfully by the Nazis in 1934, and as head of the Comintern thereafter. He kept his position as Bulgarian Prime Minister during the Tito-Stalin split, despite his advocacy with Tito of a Balkan customs union and federation just a few months before the dispute erupted (see Chapter 5). Traicho Kostov, one of his logical successors and the party leader most responsible for economic policy since 1944, did not survive the purge following the Tito-Stalin split.[2]

The actual successor, Vulko Chervenkov, had been trained in the Soviet Union for party work since his exile there in 1925. He coordinated propaganda for the Comintern from the late 1930s, and for the Bulgarian party's Central Committee after his return to the country in 1946. His background did not prepare him well, in other words, for overseeing the first Five-Year Plans. His two decades of Soviet exile did, however, prepare him to follow Stalin's lead after the split with Yugoslavia in 1948, and to reject any further delay in proceeding with rapid industrialisation and forced collectivisation according to the Soviet experience of the 1930s. In addition, Chervenkov came to power during Stalin's last years, when the Soviet Union's own reliance on propaganda slogans and the threat of arbitrary punishment reached its post-war peak. These were distinguishing features of economic policy in both countries from 1950 until Stalin died in 1953. Chervenkov had begun his regime by expelling one-fifth of a party membership that had grown to half a million. Many of those expelled, like half of the party membership, were peasants. So were many of the unknown numbers of suspected 'enemies of the people', who were sent to concentration camps in the early 1950s. All this made the atmosphere surrounding further collectivisation ominous, rather than encouraging.

The first challenges to Chervenkov's leadership none the less appeared surprisingly early in the Second Five-Year Plan (1953–7). The plan was itself a retreat from the harsh, sometimes counter-productive measures of the first. Criticism of Chervenkov for these excesses appeared in the Bulgarian Politiburo as early as 1953, and reappeared in 1955 because of continuing agricultural problems. Khrushchev's 1956 speech exposing Stalin's 'mistakes' and his 'cult of personality' was perhaps the most important, but not the first step in Chervenkov's demotion. Todor Zhivkov emerged from the new generation of post-war party leaders to become First Secretary in 1954, at the age of 43, and Deputy Prime Minister in 1956. But Chervenkov was to remain the other Deputy Prime Minister until 1961. Anton Iugov, the Interior Minister during the mass arrests of 1949–50 and one of the older generation of 'home Communists', had re-emerged in 1956 as Prime Minister. Zhivkov strengthened his position in this triumverate as the 1950s drew to a close. The influence of the other two still remained to be reckoned with until the shortcomings of the Third Five-Year Plan had become clear. Bulgarian economic policy did not therefore pass fully into the hands of Zhivkov and his post-war generation until the 1960s.

The Rise of Heavy Industry

Bulgaria's First Five-Year Plan made two important contributions to the pattern of subsequent economic development. The first was to create an institutional apparatus for long-term planning of industrial production, which had been entirely removed from private hands since December 1947. The state's existing economic ministries were subdivided into one ministry for virtually every branch of production. By January 1948, a separate and politically powerful State Planning Commission (DPK) had come into being.[3]

The long pre-war experience of the Main Directorate for Statistics (GDS), probably the most rigorous and comprehensive such agency among the Balkan states, served this new commission well. By March 1948, representatives of the new DPK and the old GDS had agreed on how to calculate national income, and by October had set out eleven criteria for calculating plan fulfilment. The major priority was to promote the rapid growth of heavy industry, maximising output and minimising inputs. This 'central place' for heavy industry was the first plan's other legacy to Bulgarian development. No programme of any pre-war regime, including the rearmament of the late 1930s, had made such a commitment.

The Communist commitment to what Soviet economists call category A, or producers' goods, could already be seen in the tentative two-year plan for 1947–8. Drawn up in 1946, well ahead of the nationalisation of a majority of existing industrial capacity, the plan was intended to restore production to the pre-war level.[4] Yet it also placed 45 per cent of state economic investment in a series of 53 new projects, 17 of them for the production of industrial outputs. Huge cement and chemical plants were to be the centrepieces of the new industrial city of Dimitrovgrad, situated on the River Maritsa in south-western Bulgaria. For the central region, a hydroelectric dam was projected for Kazanluk. This and other projects to increase electric capacity followed in the footsteps of the early Soviet emphasis on planning the expansion of electric power through GOERLO in the 1920s, well before its First Five-Year Plan of 1928–32. Expanding Bulgarian capacity also made good economic sense, given the limited construction of the inter-war period (see Chapter 3). What did not make sense was the tiny proportion of new investment, just 6 per cent, allocated to agriculture. This approach replicated Soviet investment strategy during the 1930s.

The new projects of the 1947–8 plan faced not only serious organisational and technical problems, but also severe shortages of imported or agricultural inputs. These difficulties were already clear by the time the party convened its Fifth Congress in December 1948. The occasion marked the announcement of the targets for the First Five-Year Plan, projected for 1949–53. Agriculture was promised more than before, 17 per cent of new economic investment, and industry 47 per cent. Electric power and chemical production alone were to receive 45 per cent of the industrial total. Gross industrial output was to more than double, an increase of

119 per cent, primarily on the basis of a massive 220 per cent increment for heavy industry. Light industry would increase its production by 75 per cent, and agriculture by 59 per cent. The rapid collectivisation and mechanisation of agriculture was relied upon to achieve this last target, while also freeing more labour for industry, construction and transport.

In 1952 the plan was declared fulfilled a year ahead of schedule, although annual data suggest increases averaging 80 per cent of target value.[5] Overall industrial output had climbed at an impressive 21 per cent a year, according to the somewhat questionable official statistics for 1949–52 in Table 6.1. Electric capacity and steel production had supposedly risen threefold, hard coal 3.7 times, and lignite, previously little mined, 26 times. Even if fully achieved, these impressive increases came from too small a base to provide the supplies that were needed in the mushrooming machinery sector and throughout the rest of the economy. The national network of electric power remained incomplete, the quality of electrical manufactures very low. Despite the new Soviet and Czech equipment brought in to mechanise the coal mines, their output lagged behind the planned figure. As a result, railway transport, which was just recovering from war losses of rolling stock, and factory power suffered, particularly during the winter. Here were the key 'bottlenecks' in the interrelated matrix of material inputs and outputs that had also proved a weak point for early Soviet central planning.

Agriculture received less new investment than planned, only 13 per cent of the total. Related industries suffered because agricultural targets were not met. The sector grew by only 11 per cent, compared with the planned 59 per cent. For the economy as a whole, the investment level was excessively high. In Marxist terms, the rate of accumulation (investment in fixed capital net of depreciation, but including net inventories as a fraction of net material product) reached 28 per cent by 1951–2. Like its neighbours in Eastern Europe, Bulgaria thus recorded an accumulation rate twice the contemporary West European average and 2.5 times its own 1929–39 rate of 12 per cent.[6] Huge investments in heavy industry were largely responsible.

Labour proved more difficult than capital to accumulate. A curious combination of revolutionary enthusiasm and police pressure generally prevented strikes, with the major exception of a stoppage by Plovdiv tobacco workers in 1953. But excessive labour

Table 6.1: Growth and Origin of National Product, 1939–80

A. Origin of Net Material Product

	1939	1948	1952	1956	1960	1970	1980
Net material product (NMP) per capita	100	89	121	149	226	440	825
Percentage of which:							
Industry	15	23	29	37	48	55	57
Agriculture	65	59	40	32	27	17	11
Construction	3	4	7	8	7	9	9
Trade and transport	14	10	19	17	15	16	20

B. Annual Growth (percentages)

	1949–52	1953–7	1958–60
NMP	8.4	7.8	11.6
Industry	20.7	12.7	16.2
Agriculture	−0.9	4.9	6.6
Construction	19.6	7.1	20.8

Sources: *Statisticheski godishnik na NR Bulgariia, 1982* (Sofia, 1983), pp. 12–13; *Statisticheski spravochnik na NR Bulgariia, 1984* (Sofia, 1984), pp. 14–15.

turnover could not be prevented, as it could not during the first Soviet plans. Established facilities like the Pernik coal mines recorded 100 per cent rates of turnover in 1952.[7]

Despite its desire to press ahead with heavy industrial growth at the fastest possible pace, the Chervenkov regime was forced by these problems to cut back the A targets for the Second Five-Year Plan (1953–7). Its publication in January 1953 predated Stalin's death, unlike other second plans elsewhere in Eastern Europe. Yet it made the same adjustments that are usually associated with the post-Stalin period. Total investments were to double, led by extraction of coal and iron ore, but both agriculture and light industry were to receive greater increases than heavy industry, as were housing and education. Gross industrial production was to rise 60 per cent by 1957, in 1952 prices. However, crop and livestock production was to increase more, by 66 per cent. A 40 per cent rise in personal incomes was planned. The rate of accumulation was therefore expected to fall by one-half by 1956 to only 14 per cent, as indeed it did.[8]

Industrial growth slowed with these adjustments from 21 per cent to 13 per cent a year between 1949–52 and 1953–7 (see Table

6.1). Here was the first of the surprisingly large swings that have characterised the pace of Bulgarian development throughout the post-war period. Recent Bulgarian scholarship has blamed this first swing on the slower growth of electrical power capacity and a shortage of industrial inputs, especially imports.[9] But Bulgarian coal production, however much its quantity might increase, remained of a quality too low for metallurgical coking. The Soviet Union filled some of the gap and provided finished metals, until a 1955 cut in these imports trimmed them back close to the 1950 level. Machine production for agriculture and industry combined was still only 11 per cent of manufactured goods by 1956, trailing food and textiles with 33 per cent and 14 per cent respectively. Food and textile exports to the USSR had risen in 1953–4, before being reduced in 1955 because of Soviet demands for industrial inputs instead. At least the industrial targets for this Second Five-Year Plan were more modest and were more nearly met. In addition, in 1956 the industrial share of the net material product exceeded that of agriculture for the first time, as noted in Table 6.1. Investment in consumer, or B goods, fell to 16 per cent of the industrial total by 1956, however, in the face of party pressure to reduce overall accumulation and ministerial pressure to maintain the growth of investment in heavy industry, or A goods.

Here we should acknowledge the obstacles in all these initial plans to using inputs efficiently and to producing outputs of high quality that do not appear in quantitative calculations, however refined and revised, of investment and output. Prices for Bulgarian food exports were set too low to encourage specialisation, even if investment funds were available. Imported machinery was typically outdated. Repair facilities were often primitive or non-existent. Most industrial workers and managers were poorly trained and inexperienced. The industrial labour force had doubled between 1948 and 1957, increasing from 210,000 to 450,000. Industrial management rose from 52,000 to 111,000. In addition, the attention that managers were required to give to the political education of the workers probably detracted from the economic efficiency of both. This was, in any case, an era when party credentials often counted more than performance for rapid promotion. Incentives for quality control, even of consumer goods, were minimal.

Other difficulties must have come from the rapid merger in 1948 of so many small, previously private firms into a much smaller number of state enterprises. For textiles, the total number was trimmed from 564 to 105, for metallurgy from 167 to 27 and for food processing from 746 to 77. At the same time, what had been the country with the smallest average-size for industrial enterprises in the Balkans, 24 employees in 1946, very quickly became one of the largest. By 1960, the average size of Bulgarian enterprises had reached 372. Some 6,000 firms had been reduced to 1,650.[10]

Just how this was accomplished and what its consequences were have never been properly examined. But the rapid merger of so many previously separate units, especially when joined with the rapid influx of peasant labour, suggests some diseconomy of scale. The apparently larger production of newly built plants, whether for chemicals, metallurgy, or whatever, offers indirect confirmation of this notion. Merged facilities of existing units in machine manufacture typically recorded production runs that were too short to prevent unexpectedly high costs. In textiles, mechanisation of merged enterprises was slow, electric power not self-generated, and wages too low to prevent high turnover. Case-studies of both new and merged enterprises are needed if our understanding is to proceed beyond its present state: Bulgarian celebration of the increasing quantity of output and Western criticism of inefficient use of inputs and of the poor quality of finished products.

Completing Collectivisation

A major prerequisite for the unprecedented targets of the Third Five-Year Plan was the full collectivisation of agriculture. Bulgarian economic historians now recognise the attendant abuses, but still regard its achievement during the period 1948–58 as the most important single success of that decade.

The first two phases of the drive, one voluntary and one forced, had collectivised 11.3 per cent of arable land by 1949. Two related measures had been tried, but had failed to force many peasants into the TKZS collectives during 1948. First, the Ministry of Agriculture initiated the obligatory sale to the state of all agricultural machinery and larger pieces of equipment that were privately owned. Second, the regime's 'anti-kulak' campaign against the larger peasant smallholders became much more intensive. Criticised in the press from 1944 to 1947 for individual cases of speculating or black marketeering, these larger landowners were now chastised for the size of their holdings alone. The influential party magazine *Novo Vreme* applied the Soviet pejorative 'kulak', or 'fist' (in the sense of having some hold over lesser peasants), to all with holdings over 10 hectares. This definition included some 9 per cent of peasant households and 26 per cent of cultivated land.[11] These households, however, were responsible for half of the grain marketed in 1947 and owned over 60 per cent of the tractors in operation. The state's obligatory purchase of all tractors thus hurt the larger holdings. So did the very low prices paid for grain deliveries to the state.

By 1949, the larger smallholders were reacting to adversity as Bulgarian peasants had done since the decade before the First World War. They began to sell only to the free market, consuming the rest of their crops, or they seeded less of their land. With tractors sold off, they accumulated more draft animals. They also began to barter grain to poorer peasants in return for manufactured goods or their labour. All these tactics aided survival in the short run, but they also provided the country's Communist leadership with ammunition to use against the private agricultural sector in the long run. The definition of 'kulak' was broadened again to include holdings of 4–5 hectares. New measures were sought to withhold manufactured goods or state services even from these smallholders.[12]

Such was the background to the third phase of the regime's collectivisation drive, during 1950–2. It began with the promulgation of a model charter for the organisation of collective farms. As elsewhere in Eastern Europe, these statutes were the culmination of the renewed Communist interest in collectivisation since 1950. This interest may be traced to more than political desire to tame Bulgarian smallholders and the easing of the immediate post-war food shortage. In addition, Stalin's paranoid suspicion of any further deviations, after the split with Tito in 1948, created competition between all Eastern European parties to show their devotion to Soviet first principles.

The Bulgarian statutes were closely based on the Soviet charter of 1935 in most respects. Members owed the farm a minimum number of labour-day units and might receive up to half of their projected cash income from the collective in advance. Private plots of 0.2–0.5 hectares and ownership of small, specified numbers of livestock were allowed. Two departures from the Soviet charter are worth noting. They are the member's residual ownership of his share of the collective's land (from 1958, of its assets), and his right, again until 1958, to receive rent for the farm's use of his land.[13]

The collectives' share of arable land rose spectacularly during 1950 to reach 44 per cent by the year's end. Another spurt during 1952, the last year of the First Five-Year Plan, brought the share to 61 per cent. The pace was the fastest in Eastern Europe.

The 1950 statutes themselves cannot fairly be said to have attracted many voluntary members by so-called 'self-dekulakisation'. Western analysis has emphasised, and recent Bulgarian scholarship has recognised, the role of 'extra-legal' pressures in the enlargement of membership lists. In plainer language, this meant threats, physical beatings and arbitrary arrests. Several important legal inducements must also be acknowledged. Most obvious was the continued growth of Soviet-style machine-tractor stations (MTS), rising from 71 in 1949 to 200 by 1957. Tractors per 1,000 hectares more than doubled to reach 3.6 for 1953–7, over 10 times the 1939 level, if still just one-third the Western European average. The machine-tractor stations had received monopoly rights to such modern equipment in 1948. More important, delivery quotas per hectare for the state-controlled system of agricultural marketing, already in place since the Second World War (see Chapter 4), were set higher for private holdings than for collective farms. From April 1950, holdings over 10 hectares owed the state 75 per cent of their grain crop. New income tax rates in May also hit the individual smallholder harder.[14]

Finally, the reliance of all peasants on sales to the residual free market, briefly abolished and then reinstated in 1949, was discouraged in two ways. One was the threat of criminal charges for any peasant accused of diverting produce due for delivery quotas into the free market. The other involved some recognition, even by the early 1950s, that the Bulgarian peasant would respond rather quickly to price incentives. The 1949 party plenum had frankly admitted Communist failure to win the support of the larger smallholders. It recognised the need to raise agricultural delivery prices, especially in view of the high prices of manufactured goods. Agricultural prices were soon increased for collective farms, although not enough to bring peasant income close to that of industrial labour.[15]

These legal devices were more important in the final phase of the collectivisation drive. They were combined with a 10–50 per cent reduction in the amount of compulsory deliveries after 1953 and an effort to use old Agrarian Party members to persuade the remaining peasants to join. As a result, the process was finally completed between 1956 and 1958. The land belonging to collective farms reached 77 per cent of the arable total by the end of 1956 and 92 per cent by 1958.[16] By 1956, only 756 private holdings exceeded 10 hectares. Most of those were in mountainous areas. The 3,200 collectives dwarfed the 86 state farms run on the Soviet *sovkhoz* model by the same preponderance in area as in numbers. This was the smallest share for state versus collective farms in Eastern Europe, and was very far behind the Soviet proportion of one-third. Like the rest of Eastern Europe, however, the size of the average Bulgarian collective farm, about 1,000 hectares, was much smaller, under one-fifth the average Soviet size.[17] That discrepancy was about to be repaired under the Third Five-Year Plan.

Reappraising the Great Leap Forward

Bulgaria's Third Five-Year Plan (1958–60) has invariably been indentified by Western economists as a 'Chinese-style Great Leap Forward'. The merger of the existing collective farms into 957 giant units, averaging about 4,500 hectares a piece, was regarded as a lasting monument to the leap, until still larger agro-industrial complexes were assembled in the 1970s (see Chapter 9). Yet the role of the Chinese example in the origins and implementation of the third plan is less important than Soviet priorities and a series of internal Bulgarian pressures.

The third plan's major goals had already been laid down at the Central Committee plenum of April 1956. This was too early for there to be any question of a Chinese example. The plenum was instead concerned with correcting the failings of the second plan, not only the arbitrary centralism of Chervenkov, but also the lagging production of output and especially inputs for heavy industry. Stipulations for a second large complex for ferrous metallurgy at Kremikovtsi were set down, and the whole plan was elaborated in more detail than before. The seventh party congress in June 1958 merely made the plan public.

Throughout the period 1956–8 the major internal pressure was, of all things, urban unemployment. The completion of collectivisation had led a large number of male peasants to abandon the countryside for the major towns. Once there, they needed factory jobs, as did artisans from the now-extinct private sector. (Some 3,300 artisan shops had been nationalised in 1951.) A reduction in the burgeoning state bureaucracy created new candidates for positions in factory management. The aforementioned cutback in Soviet imports for 1955 created some excess capacity in heavy industry. At the start of 1956, registered unemployment in Sofia reached 27,000 and in Plovdiv 12,000. In Plovdiv and several other towns, tobacco workers displaced by long-overdue mechanisation numbered 77,000. All jobs seekers combined may have totalled 350,000 by 1958.[18] One reason for the plan's unprecedented growth targets was the obvious need to employ these people. The original plan therefore projected the creation of 140,000 new jobs by 1959 and 400,000 by 1962. Some 10,000 agricultural workers were scheduled to go to the Soviet Union for special projects, the first large-scale movement of Bulgarian labour since the Second World War.

Foreign trade with the USSR, Eastern Europe and also the FRG largely explained the plan's new investment priorities, along with the continuing commitment to heavy industry. Food processing and agriculture *per se* were tagged for increased growth. These were areas of special Bulgarian responsibility, together with chemical fertilisers and small electrical equipment, in the now active plans of the Council for Mutual Economic Assistance (CMEA) for greater East European trade.

Created at Soviet initiative in 1949, the organisation had done little during the first two Five-Year Plans. This had initially been an era in Eastern Europe of emulating the autarkic Soviet growth strategy of the 1930s. What foreign trade occurred was handled through bilateral agreements, also familiar from the 1930s. The first CMEA sessions in 1949–50 had largely ignored Bulgaria, except to approve construction of a Romanian-Bulgarian bridge across the Danube, and to ratify the creation of the joint Soviet-Bulgarian company Gorubso (one of five joint companies) for geological surveying and mining of new deposits of petroleum, iron and non-ferrous ores. (None were discovered in the expected quantity or quality.) No more sessions were even held until 1954, after Stalin's death. By then budget funding of Gorubso surveys had already been cut back, and a Soviet scheme to expand Bulgarian cultivation of cotton, thus eliminating the need for large imports from the USSR, had been set aside. Gorubso was soon left entirely in Bulgarian hands, as were the other joint companies for housing and ship construction, for the TABSO airline and finally, in 1956, for uranium mining. Subsequent Soviet readiness to breathe some life into CMEA may be seen in the multilateral commissions and expanded trade plans put in place at the 1956 session in East Berlin.[19] After a reduction in 1955, Bulgaria now faced greatly increased export obligations to the USSR, Czechoslovakia and the GDR for the late 1950s. From the last two, there was the promise of badly needed machine imports in return.

This need had already led Bulgarian authorities to increase trade with West Germany significantly, despite the breach in formal diplomatic relations since 1957. By 1959, West Germany accounted for 8 per cent of Bulgarian import value. Table 6.2 reflects the rise in overall foreign trade that had begun by this time, albeit without a significant shift in the structure of exports or imports. Table 6.3 recalls the overwhelming predominance of Soviet trade throughout the period.

The need to increase both industrial employment and agricultural exports helped push the party leadership into its decision — taken in October 1958 — to fulfil the third plan, like the first, within three or four years. In addition, the targets for agricultural production and processed food were more than doubled. They both surpassed the 77 per cent increase planned for heavy industrial production. Such targets would presumably quadruple the modest annual growth of 2.4 per cent for agriculture that had so troubled the leadership at the party congress in June.

These decisions had already been reached by the time a party delegation, headed by Vulko Chervenkov, departed to visit China. Chervenkov returned in November full of praise for the huge Chinese farms, which relied not only on peasant labour, but also on urban party members to bring in the harvest. Perhaps as a way of reasserting his authority against the ascending Zhivkov faction within the party, he took the lead in a brief campaign to follow the Chinese example specifically. Party functionaries vied to pledge publicly that they would contribute 30–40 days of agricultural labour a year. (The Agrarian leader Stamboliiski had wanted to oblige all urban officials and professionals to do the same in the early 1920s.) In December, party leaders in Botevgrad district proclaimed the entire area a 'giant commune' on the Chinese model. And then, before the year's end, all talk of emulating China stopped as quickly as it had started.[20] The actual programmes that had characterised the Chinese experience were in any case designed

152 The First Five-Year Plans

Table 6.2: Balance and Structure of Foreign Trade, 1950–60 (in million leva)

Commodity	1950		1955		1960	
	X	M	X	M	X	M
Ferrous metals	0.5	29.3	17.4	36.1	23.9	112.1
Machinery	0.0	56.6	7.2	150.7	89.0	323.5
Chemicals	3.3	18.6	11.0	26.1	25.1	65.7
Textiles	0.8	12.5	32.4	15.1	82.7	72.1
Processed tobacco	43.4	0.0	47.1	0.0	100.5	0.5
Other processed food	39.4	1.4	58.1	4.3	142.4	25.8
Unprocessed crops	26.8	4.3	31.0	9.4	86.5	19.1
Other	22.8	32.3	72.2	50.7	118.5	121.3
Total	137.0	155.0	276.4	292.4	668.6	740.1

X = Exports; M = Imports.
Source: *Vunshna turgoviia na NR Bulgariia, 1939–1975* (Sofia, 1976), pp. 18–28.

Table 6.3: Direction of Foreign Trade, 1950–60 (percentage)

Country	1950		1955		1960	
	X	M	X	M	X	M
USSR	54.4	50.2	50.5	47.5	53.8	52.4
Czechoslovakia	14.7	15.9	10.8	16.6	9.6	9.8
Poland	10.1	9.5	3.0	4.2	3.6	3.4
GDR	5.5	3.8	13.7	9.2	9.8	11.1
Other CMEA	7.2	6.2	9.2	9.9	4.8	3.8
FRG	0.7	3.4	2.3	2.5	3.5	5.9
Other Western	7.2	1.3	8.3	9.3	12.5	13.7

Source: *Vunshna turgoviia na NR Bulgariia, 1939–1975*, pp. 18–28.

to relieve population pressure. This was hardly a Bulgarian problem by the 1950s.

Zhivkov's leadership had apparently asserted itself. The mergers of collective farms into units, which he had opposed in 1957, were in fact carried to completion from November 1958 into 1959. But their size no longer exceeded Soviet proportions, as the Botevgrad commune would have done, if it had actually been implemented. Their advantages were now publicised in Soviet terms of commitment to industrial-size units of production and to improved leverage for the apparatus of central planning in Sofia. The mystique of large-scale, mechanised production, after all, had pervaded Soviet economic thinking from the earliest days. By 1958–9, moreover, Soviet-style problems of localism, or *mestnichestvo* (the failure to co-ordinate the activities of many local enterprises enough to prevent bottlenecks and input hoarding), were being discussed in Bulgarian economic journals.[21] The logic of the larger collective farms was, in other words, allowed to stand. When less than three-quarters of the ambitious growth foreseen in the revised Third Five-Year Plan was actually achieved, however, the blame for this failure was laid on Chervenkov and his supporters. His ouster from the Central Committee and even the party followed in 1962.

During the period 1948–60, the Bulgarian economy had none the less undergone major structural changes. Industry's share of the net material product had increased from 23 per cent to 48 per cent. Agriculture's share had fallen from 59 per cent to 27 per cent. As may be seen in Table 6.1, this was the major sectoral shift of the post-war period. At the same time, crop production had risen by 1.6 per cent a year since 1950, more than the 1.4 per cent recorded for 1925–39 with a growing supply of peasant labour.[22] The completion of collectivisation had shifted 678,000 peasants — about one-fifth of the active labour force — into industrial jobs, mainly in the larger towns. The average annual increase in industrial employment peaked at 11.5 per cent during the period 1955–60, the highest rate ever recorded in post-war Eastern Europe. Some 20,000 engineers and 5,000 'economists', mainly accountants and managers, had graduated from a university system organised around science and technology as much as Leninist ideology. By the end of the period, the value produced by heavy industry matched that by light industry. Machine-building and chemical processing had begun to determine the character of Bulgarian industrial development. Food processing, especially for export, was also growing rapidly. Budget expenditures consisted largely of reinvestment in these fastest-growing sectors. Freight tonnage on a slightly increased rail network had quadrupled between 1948 and 1960, as more rolling stock was added.[23] At the same time, energy shortages persisted, and a labour shortage threatened once the peasant movement to the towns had been completed.[24]

In facing these and other problems since 1960, Bulgarian economic policy has not abandoned its readiness to create even larger-scale units of production in industry as well as agriculture. But the goal of this subsequent policy has been greater efficiency and technological modernisation, not rapid growth at any cost. Economic incentives have increasingly replaced reliance on mass mobilisation and political enthusiasm. The era of economic reform and the pursuit of intensive growth was about to begin. The era of economic revolution and extensive growth had come to an end.

Notes

1. For an overview of the Soviet transition to central planning, see Roger Munting, *The Economic Development of the USSR* (New York: St Martin's Press, 1982), pp. 63–110. The most detailed Western study is Eugene Zaleski, *Stalinist Planning and Economic Growth, 1932–1952* (Chapel Hill, North Carolina: North Carolina University Press, 1980).

2. He was tried and executed in 1949, on charges of Titoist ties and even links to Western intelligence services, which party historians have long since repudiated as false. His vulnerability derived mainly from being a 'home Communist' who had not spent long years in the USSR. Kostov was also an easy scapegoat for the economic problems of the immediate post-war period.

The other logical successor to Dimitrov was Vasil Kolarov, a 'Moscovite' in Western parlance, but also Bulgaria's able representative at the Paris peace talks in 1946. He was, however, 72 years old when he assumed party leadership in mid-1949. Kolarov himself died six months after Dimitrov.

The internal dynamics of Bulgarian political history during the late 1940s and the 1950s are not thoroughly or dispassionately treated anywhere. A brief, but useful Bulgarian account is Mito Isusov, 'The Socialist Revolution in Bulgaria', *Bulgarian Historical Review*, vol. 1–2 (1981), pp. 18–24. The most extensive Western treatments are Nissan Oren, *Revolution Administered: Agrarianism and Communism in Bulgaria* (Baltimore, Maryland: Johns Hopkins University Press, 1973), pp. 103–53; and J.F. Brown, *Bulgaria under Communist Rule* (New York: Praeger, 1970), pp. 3–39, 53–142.

3. Peter Shapkarev, *Statistiko-ikonomicheski etiudi vurkhu narodnoto stopanstvo na NR Bulgariia* [Statistical-Economic Studies of the National Economy of the PR Bulgaria] (Varna, 1982), pp. 189–229, provides a first-hand account of the commission's work on the initial long-term plan and co-ordination with the state statistical service.

4. L. Radulov (ed.), *Sotsialno-ikonomicheska politika na Bulgarskata durzhava, 681–1981* [The Socio-Economic Policy of the Bulgarian State, 681–1981] (Sofia, 1981), pp. 343–7.

5. L.A.D. Dellin, *Bulgaria* (New York: Praeger, 1957), pp. 315–17. Brown, *Bulgaria*, pp. 19–27; Nikolai Todorov et al., *Stopanska istoriia na Bulgariia, 681–1981* [Economic History of Bulgaria, 681–1981] (Sofia, 1981), pp. 436–57.

6. Zl. Zlatev, 'Socialist Industrialization in Bulgaria', in M. Isusov (ed.), *Problems of the Transition from Capitalism to Socialism* (Sofia, 1975), pp. 176–87.

7. Brown, *Bulgaria*, pp. 28–30.

8. Ibid., pp. 39–45.

9. Zlatev, 'Socialist Industrialization', pp. 183–4; G. Nikova, 'Bulgaria and Cooperation of the CMEA Member-Countries in Production Planning 1949–1956', *Bulgarian Historical Review*, vol. 12, no. 2 (1984), p. 13. Todorov et al., *Stopanska istoriia na Bulgariia*, pp. 458–60; Shapkarev, *Statistiko-ikonomicheski etiudi*, pp. 130–5.

10. S.D. Zagoroff, *The Economy of Bulgaria* (Washington, DC: Council for Economic and Industry Research, 1955), p. 21; Zlatev, 'Socialist Industrialization', pp. 198–9.

11. Vl. Migev, 'Borbata sreshtu kulachestvoto i negovoto likvidirane v Bulgariia 1944–1958' [The Struggle against Kulakism and its Liquidation in Bulgaria, 1944–1958], in Kh. Khristov (ed.), *Iz istoriia na stopanskiia i sotsialniia zhivot v bulgarskite zemi* [From the History of Economic and Social Life in the Bulgarian Lands] (Sofia, 1984), pp. 44–9.

12. Ibid., pp. 47–65.

13. Throughout 1950 rent payments accounted for over a quarter of the collective's distributed income, before dropping below 10 per cent by 1955 and disappearing by the end of the decade. Nikola Popov (ed.), *Ikonomika na Bulgariia* [The Economy of Bulgaria], vol. II (Sofia, 1972), pp. 285–308. A precise comparison of the Soviet and Bulgarian charters is found in Karl-Eugen Wädekin, *Agrarian Policies in Communist Europe* (London: Martinus Nijhoff, 1982), pp. 72–9.

14. Wädekin, *Agrarian Policies*, pp. 31–43; Migev, 'Borbata', pp. 65–76. On tractors, other inputs and agricultural output, see Gregor Lazarchik, *Bulgarian Agricultural Production, Output, Expenses, Gross and Net Product, and Productivity, at 1968 Prices, for 1939 and 1948–1970*, OP-39 (New York: L.W. International Financial Research, 1973).

15. Todorov et al., *Stopanska istoriia na Bulgariia*, pp. 450–3, 462–9.

16. Brown, *Bulgaria*, pp. 201–8.

17. Wädekin, *Agrarian Policies*, tables 6.1–6.2, pp. 85–8; Todorov et al., *Stopanska istoriia na Bulgariia*, table 3, p. 465.

18. Brown, *Bulgaria*, pp. 83–7; Boika Vasileva, 'Problemi na urbanizatsiiata i migrationsionite protsesi v Bulgariia prez prekhodniia period' [Problems of Urbanisation and Migration Processes in Bulgaria in the Recent Past], in Khristov (ed.), *Iz istoriiata*, pp. 106–8.

19. Nikova, 'Bulgaria and CMEA Cooperation', pp. 3–18, provides a useful overview of the period 1949–56. Also see Todorov et al., *Stopanska istoriia na Bulgariia*, pp. 504–12.

20. George R. Feiwel, *Growth and Reforms in Centrally Planned Economies: The Lessons of the Bulgarian Experience* (New York: Praeger, 1977), pp. 45–8; Discussion by Mark Allen in Paul Marer and John Michael Montias (eds.), *East European Integration and East-West Trade* (Bloomington, Indiana: Indiana University Press, 1980), pp. 318–20.

21. Brown, *Bulgaria*, pp. 104–7.

22. The Bulgarian rate of crop growth for 1950–64 lagged slightly behind those of Romania and Yugoslavia, and further behind Greece's 1.8 per cent. In none of these neighbouring countries, however, was there a comparable shift from rural to urban employment. See Marvin R. Jackson, 'Agricultural Output in Southeastern Europe, 1910–1938', *ACES Bulletin*, vol. XXIV, no. 4 (1982), p. 63.

23. Railway freight of 9.8 million tonnes in 1948, compared with 7 million tonnes in 1940, had increased to 25.5 million tonnes by 1956 and 38.4 million tonnes by 1960. Total railway lines in operation had meanwhile risen only to 4,136 km in 1960, from a network of 3,732 km in 1946, already extensive as a result of 1,500 km of new lines built between 1930 and 1944. A number of existing lines were double-tracked, however, during the 1950s, and a small start was made at electrification, which covered one-fifth of all track by 1970. B.R. Mitchell, *European Historical Statistics, 1750–1970* (New York: Columbia University Press, 1978), pp. 319, 334; Todorov et al., *Stopanska istoriia na Bulgariia*, pp. 470–3, 513–17.

24. Zlatev, 'Socialist Industrialization', pp. 192–200; Feiwel, *Growth and Reforms*, tables B.52 and B.53, p. 309.

7 INDUSTRY AND AGRICULTURE SINCE 1960

The Bulgarian transition to economic priorities and institutions based on the Soviet pattern was essentially complete by the end of the 1950s. It was on this basis that economic development (modern growth supported by structural change) was under way for the first time in Bulgarian history. The concentration of investment capital and the arrival of factory labour from a newly collectivised agricultural sector were the key structural changes that sustained rapid growth of heavy industry and modern technology. This is the first of three chapters devoted to the economy's rather more distinctive course of development since 1960. Bulgarian economic development derived in part from a larger commitment to foreign trade than that of the Soviet Union or of any other Eastern European country. Since 1960, moreover, the making of Bulgarian economic policy has been marked by virtually unbroken discussion about how to improve the productivity of labour and capital. The discussion has prompted recurring reforms in the initial Soviet system of central planning and ministerial control. Chapter 8 will deal with foreign trade and Chapter 9 with internal reform.

First, the present chapter must set down the further growth and structural change of the national economy since 1960. Map 2 indicates the pattern of production during this period, as well as the range of natural resources. More attention must now be paid to statistical turning-points than in the previous chapter. Increasingly less attention need be paid to discrepancies between planned and actual growth; they become less glaring after 1960 and virtually disappear after 1980.

Comparisons across the entire post-war period suggest another important change under way in the Bulgarian economy from about 1960 onward. This has been the transition from extensive to intensive growth, more precisely from growth based on increased inputs to growth based on greater productivity per input. For labour, the transition was fuelled by massive injections of new fixed capital and proceeded more rapidly than anywhere else in Eastern Europe. For capital and other inputs, the growth of productivity has been sporadic and remains illusive. Management and technology have

Industry and Agriculture since 1960 157

Map 2: Economic Resources

not improved consistently enough to increase the efficiency with which capital in particular is used. The concentration of more and more inputs into modern industrial production, however, has continued to be the principal source of structural change in the economy.

During the past decade, the overall rate of Bulgarian economic growth has itself declined. The productivity of capital has failed to keep up with that of labour. Raw materials have become more expensive, as they have everywhere in the world. Yet the record of growth remains a remarkable one, particularly when compared to economies of similar size in Western and Eastern Europe. The discontinuities of the general European performance before and after the oil shock of 1973 do not appear in the Bulgarian case.[1] If there was no economic miracle for Bulgaria in the 1960s, neither was there a serious setback during the 1970s.

Political continuity provides part of the explanation for this relatively stable performance. By the early 1960s, as spelled out in the previous chapter, Todor Zhivkov had consolidated his position as party First Secretary and had become Prime Minister. In 1971, he exchanged the latter position, now eliminated, for the Presidency of the new State Council. Under this reorganisation, Zhivkov has retained authority over the Council of Ministers, although he is no longer its chairman. He is therefore head of state as well as head of the party. The collective leadership of the Politburo of the party's Central Committee and the 27 members of the slightly larger Council of Ministers have none the less come to play the wider role in making decisions that the equivalent bodies do in the Soviet Union. Enough younger members have entered the Politburo in recent years to lower the average age to below 60, which is significantly younger than the Soviet figure.

No independently powerful figure or likely successor to Zhivkov has yet emerged. (His daughter, Liudmila, though a member of the Politburo, was never considered his probable successor nor equal; she was none the less widely mourned at her early death in 1981.) This quarter of a century constitutes the longest period of unbroken political stability under a single leadership in modern Bulgarian history. Among a population whose historical memory of the twentieth century is dominated by uncertainty and impermanence, by brief triumphs and enduring defeats, this recent continuity must be of significance. In the rest of Eastern Europe, only Hungary has had a comparable experience.

Population Growth and Labour Shortage

Bulgaria's declining birth rate has introduced demographic stability and also stagnation. The low rate of natural increase typical of post-war Eastern Europe has prevented a population explosion of the sort that has eaten up the aggregate growth achieved by a number of Third World economies, when reduced to per-capita terms. In the short run, the limited Bulgarian increases in population have helped to push ahead of per-captia growth. The long-run consequences of this demographic trend for the supply of labour for domestic demand are, however, much less desirable.

The decline of the Bulgarian birth rate predates the post-war period. This tendency had appeared in the country's predominantly rural population by the early 1920s. As noted in Chapter 2, this was primarily a peasant response to war losses and then to post-war uncertainty. Table 7.1 shows how the decline has continued, with

Table 7.1: Patterns of Population Growth, 1900–83

	Birth rate per 1,000	Death rate per 1,000	Natural increase per 1,000	Total population (thousands)	Urban share (%)	Density (per km^2)
1900	42.2	22.5	19.7	3,716	19.8	38.9
1920	39.9	21.4	18.5	4,825	19.9	47.0
1940	22.2	13.4	8.8	6,368	23.0	61.7
1950	25.2	10.2	15.0	7,273	27.5	65.7
1960	17.8	8.1	9.7	7,906	38.0	71.4
1970	16.3	9.1	7.2	8,515	53.0	76.8
1980	14.5	11.1	3.4	8,877	62.5	80.0
1983	13.6	11.4	2.2	8,939	65.0	80.7

Sources: *Statisticheski godishnik na NR Bulgariia, 1982* (Sofia, 1983), pp. 29–34; *Statisticheski spravochnik na NR Bulgariia, 1984* (Sofia, 1984), pp. 168, 180.

few interruptions, into the 1980s. A post-war baby boom pushed the rate of births per 1,000 to 25.2 by 1950, briefly recapturing the level of the mid-1930s. The decline soon resumed; the rate fell below 20 per 1,000 by 1956 and, after a small upturn for 1968–74, was down to 13.6 by 1983. An aging population has prompted a slightly rising death rate since 1965. The rate of natural increase decreased more abruptly, from 10 per 1,000 in 1958 to only 2.2 in 1983. Bulgarian economists have begun to speak seriously about the prospect of zero population growth within the next decade. A new series of tax incentives for child-bearing households and

Table 7.2: Distribution of Labour Force, 1948–83 (percentage of active labour)

	Industry	Agriculture[a]	Construction	Transport and trade[b]	Services
1948	7.9	82.1	2.0	3.7	4.3
1956	12.9	70.5	3.3	6.0	7.2
1960	21.9	55.5	5.2	8.1	9.2
1965	26.3	45.3	7.0	10.3	10.8
1970	30.3	35.7	8.4	12.5	13.1
1975	33.5	27.5	8.0	14.6	15.7
1980	35.2	23.8	8.2	15.8	17.0
1983	36.1	21.9	8.2	15.7	17.2

[a] Includes forestry.
[b] Includes communications.

Sources: *Statisticheski godishnik na NR Bulgariia, 1971*, p. 38, *1982*, p. 106; *Statisticheski spravochnik na NR Bulgariia, 1984*, pp. 16–17.

of disincentives for the childless was introduced in 1984.

Whereas Western economists regard the restriction of output and consumer demand as the major long-run burdens of reduced population growth, their Bulgarian counterparts consider capping the supply of labour as most ominous. The last significant influx of immigrant labour came from Thrace and the Macedonian lands in the early 1920s (see Chapter 2). For the expanding industrial sector of the 1950s and 1960s, the influx of peasant labour from the countryside to the towns provided an ample supply of new workers. But by the late 1960s, just as the urban share of the population passed 50 per cent, the Bulgarian press and scholarly journals began lamenting a shortage of industrial labour. By the early 1980s, Bulgaria's urban population of working age had begun to decline in absolute numbers. Not even a slight increase is forecast until the 1990s.[2] Tables 7.1 and 7.2 indicate how much rural migration and the growth of the industrial labour force have slowed down since 1970. At 49 per cent of the industrial and overall labour force by 1981, the absorption of women has reached its upper bound. The declining birth rate has compounded these limitations, as has the reduced number of working hours. The weekly norm was cut from 48 hours to 42.5 hours with the introduction of the full weekend since 1974. By 1975, the demand for labour reportedly exceeded supply by 2 per cent in industry and by 3 per cent and 4 per cent in construction and transport. An annual 3.3 per cent increase in industrial employment for 1976–80 allowed this gap to widen, given the higher rate of industrial growth.[3]

Three consequences of this labour shortage should be borne in mind for the discussion of national and industrial production that follows. One is the pressure placed on this relatively fixed labour force to increase its productivity, and thereby permit intensive growth. Another is the planners' temptation to replace scarce labour with excessive investment of capital, running the risk of increasing the demand for labour still further, if new machinery does not actually save labour. A third is the workers' temptation to escape these pressures by moving from job to job. The annual turnover of the existing force of industrial labour was still 29 per cent in 1981, a high figure by Eastern or Western European standards, although well under the 50–100 per cent rates recorded in the 1950s.

National Income and Industrial Growth

National income, by Bulgarian definition, consists of net material production (NMP) of goods, minus that of most services, including government, all indirect taxes and depreciation. The NMP has maintained an impressive, if declining rate of growth. Official statistics record a rise of 8.3 per cent for 1966–75, before slipping 6.1 per cent for 1976–80 and to 4.1 per cent since 1980. It is difficult to compare these net aggregates of goods with the gross value of goods and services produced, or GNP, used in Western practice.

Table 7.3 matches NMP figures with the principal Western effort, by Thad Alton and associates, to calculate rates of growth for Bulgarian GNP since 1960. Alton's addition of one-fifth for services, plus his factor-cost formula for price-indexing physical output, and a higher weighting of agriculture in total output combine to reduce the Bulgarian rate of aggregate growth and especially industrial growth after 1965. The downward bias in Alton's calculations emerges from their dollar conversions, which continue to rank Bulgaria behind Romania, at $3,830 compared with $4,240 per capita for 1981. The World Bank has reversed those rankings, dramatically for 1981, with Bulgaria's $4,150 virtually equal to Hungary's $4,180, and over twice the Romanian $1,900. Estimates by the United Nations Economic Commission for

Table 7.3: Comparison of Official and Alton's Rate of Growth, 1961–83 (percentage average annual growth)

A. NMP, GNP, Industrial and Agricultural Growth

	1961–5	1966–70	1971–5	1976–80	1981–3
Net material product (official)	6.7	8.8	7.8	6.1	4.1
Gross national product (Alton)	6.7	5.1	4.7	0.9	2.9[a]
Industry (official)	11.7	10.9	9.1	6.0	4.5
Industry (Alton)	11.5	4.7	6.4	3.2	2.8[a]
Agriculture (official)	3.2	3.5	2.9	0.9	1.2
Agriculture (Alton)	0.0	0.2	2.2	−3.4	4.6[a]

B. Percentage weighting, 1975

	Industry	Agriculture	Construction	Trade and transport	Services
NMP (1971 prices)	54.0	18.6	9.0	15.4	—
GNP (1975 prices)	35.1	27.6	6.7	14.3	21.0

C. Other Official Growth Rates

	1961–5	1966–70	1971–5	1976–80	1981–3
Capital investments	7.9	12.5	8.6	4.0	6.8
Real wages	2.0	5.3	3.0	0.5	2.6
Foreign trade turnover	14.6	11.3	12.0	8.5	6.9

[a] 1981–2.

Sources: *Statisticheski spravochnik na NR Bulgariia, 1984*, pp. 14–15, 169; T. Alton et al., *Economic Growth in Eastern Europe, 1965, 1970 and 1975–1981*, OP-70 (New York, 1982), p. 7.

Europe, which used physical indicators to calculate the Eastern European countries' GDP (gross domestic product, GNP minus foreign trade, but including services), were unfortunately replaced with official NMP figures after 1973. For the period 1958–60 to 1967–9, the GDP estimates placed Bulgaria one point ahead of the Eastern (and Southern) European averages (7.4 per cent, compared with 6.5 per cent and 6.6 per cent respectively) and further ahead of Western Europe's 4.7 per cent.[4]

This sort of exercise raises serious statistical problems. The quality of production resists international comparison, as does the

Table 7.4: Sources of Non-Agricultural Growth, 1953–74 (percentages)

Aggregate growth from	1953–7	1958–60	1961–5	1966–70	1971–4
Labour force	54	68	48	30	32
Labour productivity	43	32	43	71	62
Capital stock	79	45	128	100	133
Capital productivity	−26	39	−84	−16	−46
Capital broadening[a]	68	151	38	30	27
Capital deepening[b]	−21	−24	74	70	72

[a] Defined as increasing the labour force at a constant capital/labour ratio.
[b] Defined as increasing the capital/labour ratio.

Sources: M. Allen, 'The Bulgarian Economy in the 1970s', JEC, *East European Economies Post-Helsinki* (Washington, DC: US Government Printing Office, 1977), p. 648.

role of foreign trade in domestic growth. Pricing formulas to compensate for the absence of the market mechanism are hard to standardise. For Bulgaria by itself, there is the added problem of weighting agricultural production in an economy that has undergone such a rapid transition from peasant agriculture to factory industry.[5] With the above limitations in mind, let us none the less look more closely at industry and agriculture, the two major sectors.

By 1956 the industrial share of Bulgarian net material product had exceeded that of agriculture and reached 48 per cent by 1960, according to the official figures in Table 6.1. It reached 57 per cent by 1980, though admittedly it was buoyed by overpricing. Agriculture had by then slipped from 27 per cent to 11 per cent, well below the 24 per cent represented by construction, trade and transport combined.

The Bulgarian turn toward intensive growth during the 1960s emerges clearly from a World Bank economist's calculation of the contribution made by labour, capital and material inputs to non-agricultural NMP. Table 7.4 differentiates between the growth attributable to increased amounts of these inputs compared with the growth attributable to their greater productivity. The increase in labour productivity exceeded the contribution of larger employment for the first time in the period 1961–5. The productivity of massive amounts of new capital investment and material inputs, on the other hand, continued to decline significantly except for the period 1966–70. The ratio of capital to output doubled between 1960 and 1970. Capital now deepened increasingly more than it broadened, according to Table 7.4.

Industry accounted for about two-thirds of Bulgaria's non-agricultural growth during the 1960s. The improved productivity of industrial labour, and especially capital, during the Fifth Five-Year Plan (1966–70) over the Fourth (1961–5) did not, however, translate into a higher rate of growth for industrial output. According to Table 7.3, annual growth for 1966–70 declined slightly, from 11.7 per cent for 1961–5 to a still impressive 10.9 per cent. This decline masks the more efficient use of capital and material inputs for 1966–70 suggested by Table 7.4. The reliance of official figures on Five-Year Plan periods is deceiving in this instance. It overlooks the unusually large increment for the last year of the fourth-plan period. Industrial output increased by 15 per cent in 1965, and averaged 13 per cent a year for 1965–8. Otherwise, Bulgarian industrial growth has displayed remarkable stability since 1960, increasing annually by about 10 per cent for 1961–4, 9 per cent for 1969–75 and 6 per cent for 1976–83, all without more than one point's deviation. This record may be contrasted with the relative volatility of NMP increases. Their standard deviation for the period 1953–79 was fully 4 per cent, less than the Romanian or Hungarian figures, but double the 2 per cent average for Eastern and Western Europe.[6]

The other distinctive feature of Bulgarian industrial performance during the 1960s was a still higher growth rate for net capital investment. Its average increment of 12 per cent for 1960–70 surpassed even the Romanian figure to lead all Eastern European countries, again buoyed by 23 per cent increments in 1966 and 1967.[7] Producer (or A) goods naturally led the way, receiving over four-fifths of industrial investment, while still accounting for barely one-half of output. The overall percentage of capital accumulation (investment in fixed capital net of depreciation, but including net inventories) in national income fluctuated, but averaged 29 per cent throughout the decade. As the supply of new labour began to drop, capital deepened with a vengeance. Table 7.5 reveals that the industrial share of capital investment peaked during the late 1960s.

During the 1970s, the growth of net investment for the entire economy slowed to 7 per cent a year, but continued to be erratic. So did the overall share of investment in national income. This accumulation ratio rose to 33 per cent in 1975 and then fell to 25

Table 7.5: Capital Investment and Accumulation, 1949–80 (percentages)

Fixed capital investment	1949	1956	1960	1965	1970	1975	1980
Industry	31.4	36.8	34.2	44.8	45.2	39.9	41.9
Agriculture[a]	12.4	22.9	29.7	19.7	15.8	14.7	12.4
Construction	2.2	0.5	1.6	2.7	2.9	4.1	2.5
Transport	16.5	6.2	5.4	6.1	7.8	12.0	9.7
Housing	22.9	23.8	19.2	16.9	15.8	15.3	20.2
Education, science, arts	2.9	3.6	3.5	3.0	4.2	4.9	4.9
Health, insurance, tourism	2.2	1.9	1.5	1.2	1.6	1.5	1.2
Other	9.5	4.3	4.9	5.6	6.7	7.6	7.2
Total	100.0	100.0	100.0	100.0	100.0	100.0	100.0
Ratio of accumulation to NMP used	–	14.3	27.5	28.3	30.8	32.8	25.0

[a]Including forestry.

Sources: *Statisticheski godishnik na NR Bulgariia, 1971*, p. 51, *1982*, p. 138, 146; *Statisticheski spravochnik na NR Bulgariia, 1976*, p. 14.

per cent by the end of the decade. For 1981, it climbed to 27 per cent, but slipped back to 22 per cent in 1983. Gross investment, including depreciation allowances, has grown more steadily, but unfortunately because these allowances have been used more to accumulate inventories of unsold goods and to stockpile unused inputs rather than to replace old machinery.

Rising input costs have helped to keep the productivity of capital from reversing its decline since 1970. Fuel and energy inputs now became more expensive, as in the rest of the world, but also continued to be used in larger quantities than in Western Europe. Chapter 8 will examine the predominance of imports in providing them. According to a Western estimate, the cost of all inputs in 1980 rose faster than industrial output, by 6 per cent, as against 4 per cent in real terms, to hold the real increase in industrial value-added output to 0.4 per cent.[8]

Profits on total industrial assets climbed steadily since the 1960s, from 9.4 per cent in 1965 to a peak of 20.7 per cent in 1979.[9] These profits became important by the 1970s, as enterprises increasingly relied on them for future investment funds (see Table 9.2). Not until the early 1980s, as we shall see in Chapter 9, was this reliance accompanied by greater attention to modernising existing facilities than to investing in new capacity.

Industrial Structure: Old Metallurgy and New Machinery

More readily understandable and perhaps more important than this maze of aggregate changes in production and investment have been several significant shifts in industrial structure. As would be expected, given the emphasis on investment in heavy industry, the share of food processing has declined. Textiles and other consumer goods have retained their small portions of total output. The principal gainers, as may be seen in Table 7.6, have been the branches of metallurgy, machinery, electronics and chemicals. By 1980 they accounted for 36 per cent of industrial production, compared with 17 per cent in 1952.

Table 7.6: Structure of Industrial Production, 1939–83 (percentages)

Branch	1939	1952	1960	1970	1980	1983
Electricity	1.8	2.1	2.0	2.5	2.4	3.9
Fuel, heat	4.6	3.2	2.8	4.6	3.7	1.4
Metallurgy	0.5	3.7	5.6	6.6	3.4	3.5
Machinery	2.4	10.0	12.4	16.5	15.6	14.2
Electronics					7.6	8.8
Chemicals	1.9	3.1	3.7	7.5	8.9	8.2
Food processing	51.2	39.2	33.5	25.4	22.9	26.9
Textiles	19.8	14.7	13.5	9.1	5.1	5.5
Construction, wood processing	1.8	2.2	3.1	3.7	4.9	4.5
Other	16.0	21.8	23.4	24.1	22.5	28.1
	100.0	100.0	100.0	100.0	100.0	100.0

Sources: *Statisticheski godishnik na NR Bulgariia, 1971*, p. 73; *Statisticheski spravochnik na NR Bulgariia, 1984*, p. 94.

Within heavy industry itself, two continuing trends seem to be significant. One is the enduring Bulgarian commitment to producing basic ferrous metals, however high the cost, in order to avoid dependence on imports. The second is the somewhat contradictory commitment to producing internationally competitive machinery, especially electrical equipment. During the past decade, according to a number of Western businessmen, the quality of Bulgarian electrical equipment has improved significantly. In order to examine both metal and machine production properly, however, microeconomic case-studies of individual enterprises would be required. Bulgarian official statistics have not published the necessary information for this. Native economists have considered, but not yet prepared, case-studies.

The Kremikovtsi metallurgical complex near Sofia is one exception, because of its size and troubled history. It began production in 1963, after nearly a decade of debate and planning. The complex has been called both the giant and the graveyard of Bulgarian ferrous metallurgy. The initial Lenin complex at Pernik, also near Sofia, had opened in 1953. Soviet-supported geological surveys the following year (see Chapter 6) had suggested that major new deposits of higher-quality iron ore would support a second complex. It was soon clear that such deposits did not in fact exist. Large imports of enriching ore would now be needed. The project none the less went ahead, with the USSR as the potential ore supplier. Some 80 per cent of the necessary equipment was also imported from the Soviet Union. Construction got underway in 1960. The project reportedly took *one-fifth* of total investment in Bulgarian industry for 1962–3.[10] Since opening, the complex has expanded to include four blast furnaces, three coking plants and rolling mills. Its output for the period 1963–78 accounted for over half of the national production of steel, cast and rolled iron. Bulgarian output of these goods has ranked ahead of all the other Balkan states, with the exception of Romania, on a per-capita basis.

The problems associated with the complex have been equally prodigious. The fraction of Bulgarian ore and coking coal used was never as much as half, and had dropped to less than one-quarter by the mid-1970s. Extraordinary wages, paid to restrain a high turnover rate, and delays in construction, delivery or repair forced costs upward. The enterprise has never shown a profit. Its production has consistently failed to meet planned targets, or even to use more than three-quarters of plant capacity. For Bulgarian ferrous metallurgy as a branch, labour productivity's average annual increase fell by over one-half to 3.5 per cent between 1971–5 and 1976–80, well below the 5.2 per cent recorded for industry as a whole.

The case of the Kremikovtsi complex helps us to understand the price that Bulgarian industry has paid for continuing to consume metal, fuel and other inputs at the higher, Eastern European level per unit of output, rather than the lower Western one. Bulgarian economists recognise the cost of this expensive import substitution, but do not evidence any readiness to abandon existing metallurgical capacity as 'old industry'.[11] Lower-cost and more specialised metals are expected to come neither from Kremikovtsi nor Pernik, nor from, say, Czech or Swedish imports, but rather from the country's third complex, scheduled to begin operation near Burgas in 1985. Its location by a Black Sea port at least places it much closer than Sofia, some 400 km from the coast, to the necessary imports, primarily Soviet, of iron ore and coking coal.

The branches of Bulgarian industry that have made the most progress since the 1960s in reducing their costs and in modernising their production are those most committed to exports. Goods sold outside the country have been primarily responsible for the average annual increase of 16 per cent recorded by the machine-building and electronics branches over the period 1960–80. During that period, these two branches combined to become the leading sector of Bulgarian industrial production, with 23 per cent of the 1980 total. Their foreign sales increased rapidly enough to suggest that export creation was more responsible than import substitution. Such substitution had been the pre-war stimulus for most Bulgarian industrial growth (see Chapters 2 and 3). The two branches' combined share of a rapidly rising export value jumped from 13 per cent in 1960 to 27 per cent in 1970 to 55 per cent by 1982. Leading the way in exports have been the well-known fork-lift trucks and electrical hoisting gear, produced by Balkancar, Bulgaria's largest single industrial enterprise, and a lesser known, but increasing amount and variety of computer equipment. Each now accounts for almost one-fifth of the branches' export value.[12] The same sort of export orientation may be found in certain chemical enterprises, particularly the pharmaceutical and cosmetic producer Pharmachim. Its exports, packaged and marketed with growing sophistication, helped to push chemicals to 8.2 per cent of total industry output by 1983, and nearly doubled its share of export value, to 3 per cent, between 1965 and 1983.

Processed foods, beverages and tobacco have been unable to maintain their comparably large share of total exports, slipping from 40 per cent in 1965 to 23 per cent in 1975 and 13 per cent by 1983. Their proportion of industrial output has held up better. A 32 per cent share in 1965 fell to 25 per cent by 1975 and 23 per cent by 1980, but rebounded to 27 per cent for 1982–3. Investment in new plant and equipment has none the less lagged. The branch received a meagre 1.5 per cent of total industrial investment for 1958–78 and 6 per cent since then.[13] Most processing facilities remain scattered in smaller plants with old equipment. Their modernisation has become a pressing priority for agriculture as well as for industry.

Growth of Agriculture

Bulgaria's agricultural record is an impressive one, for at least part of the post-war period. From 1956 to 1970, crop and animal production grew at an annual average of 4.1 per cent per capita, a higher rate than that achieved anywhere else in Eastern Europe.

For the decade before 1960, the advance of its final years came primarily as a result of massive new investments associated with the completion of collectivisation. From 15 per cent in 1950, agriculture's share in net Bulgarian investment rose to 28 per cent by 1960, also the highest figure for Eastern Europe. Bulgaria was thereby the first country in the area to overcome one of the most serious failings of the initial Soviet pattern of collectivisation. This was the notion that reoganisation was a substitute for large investment in the sector.[14]

The period since 1960 has witnessed Bulgarian reliance on increasing labour productivity in agriculture. For 1960–70, the average annual increment in output per person occupied in agriculture was in fact 8.6 per cent, well above rates recorded elsewhere in the region. Helping to force this intensive growth was the continuing decline in the agricultural labour force. The yearly losses of labourers had averaged 3 per cent of the total during the 1950s and 4 per cent during the 1960s. Agricultural labour had fallen over 40 per cent from its 1950 total by 1970. Its numbers now dropped below industrial labour for the first time.[15] No new investment surge appeared after 1960, however. The agricultural share of total Bulgarian investment declined rapidly from its 1960 peak to 19 per cent by 1965 and to 16 per cent by 1975. Gross investment still increased in absolute terms, at an annual rate of 4.5 per cent for 1960–79.

After 1970, however, agricultural production itself grew less rapidly, despite an average increase of 6.5 per cent in labour productivity until 1979. Output alternated between 4–5 per cent annual advances and, in years of bad weather, virtual zero growth, a pattern which had persisted into the 1980s. The combination of a spring drought and heavy summer rains in 1983 helped to force a fall in agricultural production of 7 per cent compared with that of 1982.

Table 7.7: Index of Agricultural Output per capita, 1932–81 (1932–8 = 100)

	1948–52[a]	1956–60[b]	1966–70	1971–5	1976–80	1981
Crops	97	131	181	188	191	200
Livestock	82	100	149	175	208	234
Total	91	119	169	183	199	210

Percentage share of total	1956	1960	1965	1970	1975	1980	1983
Crops		64.4	67.3	64.4	64.7	56.7	52.2	44.9
Livestock		35.6	32.7	35.6	35.3	43.3	47.8	55.9

[a] Missing 1949 and 1951.
[b] Missing 1958.
Sources: *Statisticheski godishnik na NR Bulgariia, 1982*, pp. 18–19; *Statisticheski spravochnik na NR Bulgariia, 1976*, p. 51, *1984*, p. 52.

Bad weather alone does not explain the reduced growth rates since 1970. The vulnerability to limited or irregular rainfall has always been a fact of Bulgarian agricultural life, as noted in the Introduction to this volume. The principal man-made problem since 1970 has come from costs of production that have risen without a comparable advance in yields or labour productivity. Bulgaria ranks ahead of its Balkan neighbours in the application of fertiliser per hectare, but domestic production continued throughout 1980 to lag behind consumption. Expensive imports of higher-quality mineral fertilisers fill the gap. The Bulgarian press has periodically lamented the poor results from the introduction of mechanically powered equipment and irrigation.[16] They have apparently been used too little or too irregularly after investment funds have secured them.

Perhaps one-half of the existing irrigation network is not fully used. That network has grown rapidly from only 126,000 hectares in 1950 to 715,000 by 1960 and to one million by 1968. Plans to extend irrigation to 2 million hectares, or nearly half of Bulgaria's 4.7 million hectares of cultivated land, have not materialised. The lagging productivity of irrigated land suggests one reason for a total network that, in 1983, was still only 1.2 million hectares. The irrigated area has increased yields only irregularly. Where it has, production costs have risen 40 per cent in return for a yield 20 per cent higher.

Tractor numbers (in 15 horsepower equivalents) grew by one-half during the 1960s to reach 153,000 units by 1980. This total, when divided by arable land, continues to exceed the Romanian and Soviet figures, if not the Yugoslav. An increasing number of older units none the less needs replacement. To these higher costs must be added a level of agricultural wages which doubled between 1960 and 1980. All of these costs, according to the American calculations by the Alton team, combined to increase the sum of agricultural expenses and depreciation by 116 per cent between 1975 and 1980.[17]

The Changing Balance of Crops and Livestock

Two major trends have reshaped the structure of Bulgarian agricultural production since 1960. One is the growth of livestock at the expense of crop cultivation, backed by a related rise in feed as against bread grains. The other is a shift away from the industrial crops, primarily tobacco and cotton, that had become so important by the inter-war period, in favour of fruit, vegetables and vineyard production. Both of these trends may have served, at least in part, to mitigate the reduced and erratic growth of aggregate crop production since 1975.

Raising livestock, particularly pigs, had played a smaller part in Bulgarian agriculture during the nineteenth century than in that of its neighbours. The growth of husbandry over the past 25 years has continued the post-1900 reversal of a previously modest role (see Chapter 1). Since the Second World War, as late as 1970, livestock accounted for only 35 per cent of the value of Bulgarian agricultural production. By 1983, if we use the higher 1982 prices for livestock, that proportion had risen dramatically to 55 per cent.[18] Grassland now took 24 per cent of the country's virtually stable agricultural area, compared with 17 per cent in 1965. Land under crop cultivation had thus declined absolutely.

How had this upsurge in livestock occurred? Much of the answer lies with meeting export demand and rearranging agricultural organisation, topics treated in the next two chapters. Here we must set down the structure of the upsurge. Pork and poultry have been largely responsible. Pigs provided 45 per cent of live-meat value produced in 1980, after their numbers were doubled during the 1970s. Faster reproduction and fattening techniques have proved effective, although the slaughtering weights are still low by Western European standards. Poultry production has relied more heavily

Table 7.8: Annual Average Output per capita of Agricultural Products, 1939–83

	Wheat	Corn	Tobacco	Tomatoes	Grapes	Milk	Eggs[a]	Meat
1939	317	159	6	7	104	105	117	32
1953–7	267	159	8	39	64	97	110	34
1961–5	273	198	13	91	125	148	167	48
1966–70	344	256	13	85	135	183	190	59
1971–5	370	290	15	91	122	194	204	67
1976–80	399	301	14	94	117	224	245	85
1981–3	483	383	13	91	127	263	283	92

All outputs are given in kilograms, with the exception of eggs, where the number of eggs are given.

Sources: *Statisticheski godishnik na NR Bulgariia, 1982*, pp. 18–19; *Statisticheski spravochnik na NR Bulgariia, 1984*, p. 137.

on the same industrial techniques of accelerated fattening to become the second largest source of meat. Bird numbers rose by one-third during the 1970s. Egg sales have also risen significantly, up by two-thirds for 1970–83. Again, the number of eggs laid annually per hen remains low (145) by Western European standards. The number of cattle increased to 45 per cent of the animal numbers, after a spurt during the second half of the 1970s. The source was not meat, but rather milk production, which was up 54 per cent for 1970–83.[19] Table 7.8 indicates the near tripling of milk, egg and meat production per capita that has occurred since the late 1950s.

The ascendancy of livestock has also had its effect on the structure of grain production. The quantity of feed grains harvested has, not surprisingly, increased from 92 per cent of the total for bread grains in 1965 to 136 per cent by 1978. Wheat cultivation continued a decline that had begun after the First World War (see Chapter 2). Its area fell 11 per cent for the period 1965–75. Rye dropped fully 43 per cent. Corn, barley and soya beans (all livestock feeds) more than took up the slack — in output, if not in area. Together they rose sufficiently to push up gross grain production by an average of 4.1 per cent for 1965–75. The area under barley cultivation increased by 20 per cent, while yields rose 27 per cent. The initially small soya-bean area increased threefold, and output doubled. Corn production grew less as a result of the 8 per cent increase in cultivated area between 1965 and 1975, than as a result of the use of more fertiliser, greater irrigation and the introduction of high-yield hybrids. Overall, the cereal share of total crop and vineyard land had fallen to 59 per cent by 1970, from 75 per cent in both 1938 and 1948.

Since 1975, however, the lesser vulnerability of wheat to drought compared with corn has prevented its eclipse by the latter as the leading Bulgarian grain. Wheat yields have also risen during the past decade to average 43 hectolitres for 1980–2, virtually matching the Hungarian and East German levels. They are half again those of Romania and Yugoslavia. (Grain production for 1978–80 averaged 44 per cent wheat, 32 per cent corn and 18 per cent barley.)

What has continued, however, is the rising share of animal feed, compared with food or seed, in Bulgarian grain consumption, including imports. The feed share increased from 55 per cent to 71 per cent between 1975 and 1980.[20] Such a shift makes it difficult to judge the significance of the reduced and more erratic performance of aggregate crop production. It averaged only a 2 per cent annual increase for 1975–82.

The same uncertainty accompanies the changing structure of the other main type of Bulgarian agricultural cultivation, the so-called specialty crops. The wide spread of sugar beet, sunflower, cotton and especially tobacco during the 1920s may be recalled from Chapter 2. Since 1960, these industrial crops have increasingly given way to directly edible fruit and vegetables or wine grapes.[21] Tobacco still accounted for 13 per cent of gross agricultural production during 1981–3, but its physical output and yield continued a decline unbroken since the early 1970s. Relocation of the tobacco fields in order to introduce new varieties may explain the slightly falling yields. Yields have also dropped for sugar beet. Its output stagnated in the 1970s and fell off after 1980. Sugar beet still managed to cover domestic demand, as it has done since the interwar period. Sunflower production has risen little. Reduced hectarage has largely cancelled out improved yields since 1980. Cotton cultivation has experienced the clearest decline. Its area has fallen steadily to one-sixth of the 1960 average by 1983. Yields have also slipped, despite the concentration of cultivation in the warmer, south-eastern districts. Imported Soviet cotton was already covering over 80 per cent of domestic consumption by the end of the 1940s, as noted in Chapter 5.

The relatively small rise in fruit and vegetable production since 1960 has resulted mainly from the increased area under cultivation, rather than from better yields. Most of this increase took place during the 1960s. Tomatoes and apples led the way. By the 1970s tomatoes accounted for half of the Bulgarian vegetable production, on land that was 94 per cent irrigated. Apple and other fruit orchards were by this time 90 per cent irrigated. The vineyard area remained larger than that of all orchards. Wine grapes recorded a 25 per cent increase in yields for the period 1976–82 over 1965–75, thus explaining their greater output.[22]

The sum of these changes in the structure of special crops was to boost the fruit, vegetable and vineyard share of total agricultural production slightly, from 15 per cent to 16 per cent for 1975–82. Industrial crops meanwhile declined from 15 per cent to 10 per cent for 1975–82.

A Summary of Structural Change

A balance sheet of structural change since 1960 in Bulgarian agriculture and industry should therefore reflect the following shifts. Livestock and feed grains have made the most rapid agricultural advance. They now account for two-thirds of the value of production. Overall grain cultivation has become concentrated almost exclusively in wheat, corn and barley, raising vulnerability to bad weather. The share of special crops in agriculture production declined to 26 per cent by 1982, from 30 per cent in 1965 (or in 1925). Fruit, vegetable and vineyard cultivation advanced slightly in absolute terms. The industrial manufacture of foodstuffs has depended more on meat than on fruit or vegetables in order to keep a one-quarter share of total industrial production constant since the mid-1960s.

Machine-building, electronic and chemical manufactures have led the rise of producers' goods to industrial pre-eminence, accounting for 55 per cent of output by the 1970s. Continued investment in and reliance upon high-cost domestic metallurgy has held back the otherwise improved performance of these leading branches, particularly their exports. The spread of a serious labour shortage from agriculture to industry, followed by slower rates of aggregate growth since 1975, has intensified pressure throughout the economy to use labour more intensively and capital more efficiently. With the possible exception of the late 1960s, less progress has been made on the latter task.

Notes

1. A good recent overview of the post-war Western European experience before and after 1973 is Angus Maddison, *Phases of Capitalist Development* (Oxford: Oxford University Press, 1982).
2. L. Radulov, *Naselenie i ikonomika* [Polulation and Economics] (Sofia, 1981), pp. 58–112, furnishes a frank Bulgarian appraisal of past trends and future prospects.
3. Mark Allen, 'The Bulgarian Economy in the 1970s', in Joint Economic Committee of the US Congress (JEC), *East European Economies Post-Helsinki* (Washington, DC: US Government Printing Office, 1977), pp. 662–4; Marvin R. Jackson, 'Bulgaria's Economy in the 1970s: Adjusting Productivity to Structure', in JEC, *East European Economic Assessment*, pt I (Washington, DC: US Government Printing Office, 1981), pp. 555–8.
4. United Nations Economic Commission for Europe, *Economic Survey of Europe, 1971*, pt I (New York: UN Secretariat, 1972), pp. 6–8; Thad Alton et al., *Economic Growth in Eastern Europe, 1965, 1970, 1975–81*, OP-70 (New York: L.W. International Financial Research, 1982), p. 22; *European Marketing Data and Statistics, 1983* (London: Euromonitor Publications, 1983), p. 167. A good critical review of available estimates is Paul Marer, 'Economic Performance and Prospects in Eastern Europe', JEC, *East European Assessment*, pt I (Washington, DC: US Government Printing Office, 1981), pp. 87–91.
5. This is the view of Bulgaria's leading expert on national accounts, Peter Shapkarev, in his *Statistiko-ikonomicheski etiudi vurkhu narodnoto stopanstvo na NR Bulgariia* [Statistical-Economic Studies on the National Economy of the PR Bulgaria] (Varna, 1982), pp. 127–32.
6. Nita Watts, 'Eastern and Western Europe', in Andrea Boltho (ed.), *The European Economy: Growth and Crisis* (Oxford: Oxford University Press, 1982), table 9.4, p. 265. Watts's article (pp. 259–86) affords a balanced and useful comparison of post-war Eastern and Western growth records.
7. Nikolai Todorov et al., *Stopanska istoriia na Bulgaria, 681–1981* [Economic History of Bulgaria, 681–1981] (Sofia, 1981), 493–503; George R. Feiwel, *Growth and Reforms in Centrally Planned Economies: The Lessons of the Bulgarian Experience* (New York: Praeger, 1977), pp. 30–7, 271.
8. Marvin R. Jackson, 'Recent Economic Performance and Policy in Bulgaria', in JEC, *East European Economies: Slow Growth in the 1980s*, pt 3 (Washington, DC: US Government Printing Office, 1985), pp. 6–8.
9. Ibid.: Todorov et al., *Stopanska istoriia na Bulgariia*, pp. 493–4, 540; *Statisticheski godishnik na NR Bulgariia*, [Statistical Yearbook of the PR Bulgaria], *1982* (Sofia, 1983), p. 148.
10. J.F. Brown, *Bulgaria under Communist Rule* (New York: Praeger, 1970), pp. 156–60: Radio Free Europe Research (RFER), *Bulgarian Situation Report*, 20, 4 December 1978; *Statisticheski godishnik na NR Bulgariia, 1982*, p. 638; *Ikonomicheski zhivot*, 3 March, 1976, and 20 September, 1978.
11. See, for instance, the article by I. Dimitrov in *Ikonomicheski zhivot*, 16 March 1983, as reported in RFER, *Bulgarian Situation Report*. 5. 18 April 1983, p. 15. The author received the same unanimous opinion in interviews with faculty members of the Higher Economics Institute Karl Marx in Sofia in May 1984. For Western analysis of the higher (8–27 per cent) cost of Bulgarian output per input, compared with Western European output per input, and of the continued underultilisation of machinery because of scarce metal inputs, see Feiwel, *Growth and Reforms*, pp. 177–8; and also Feiwel, 'Economic Development and Planning in Bulgaria in the 1970s', in Alec Nove, Hans Hermann Höhemann, Gertrud

Seidenstecher (eds.), *The East European Economies in the 1970s* (London: Butterworths, 1982), pp. 222−3.

12. *Bulgaria's Presence in 5 Continents*, 11/101 (Sofia Press, 1983). Balkancar employs 45,000 people, and is the world's largest producer of fork-lift trucks. Its hard-currency surplus for 1983 topped $100 million and allowed the firm to bid, albeit unsuccessfully, for an equity share in a French enterprise. See David Buchan, 'Balkancar's European Ambitions', *Financial Times* 7 June 1984, p. 24.

13. OECD, *Prospects for Agricultural Production and Trade in Eastern Europe*, vol. 2, *Bulgaria* (Paris: OECD, 1982), pp. 188−91.

14. On Soviet collectivisation and agricultural performance under Stalin and Khrushchev, see Karl-Eugen Wädekin, *Agrarian Policies in Communist Europe* (London: Martinus Nijhoff, 1982), pp. 14−30, 44−62.

15. Gur Ofer, 'Growth Strategy, Specialization in Agriculture, and Trade: Bulgaria and Eastern Europe', in Paul Marer and John Michael Montias (eds.), *East European Integration and East-West Trade* (Bloomington, Indiana: Indiana University Press, 1980), pp. 308−9. Bulgaria's agricultural labour force per hectare fell from 38 to 17 for 1960−78, but still remained high by Western European standards. OECD, *Prospects*, vol. 2, *Bulgaria*, pp. 157−8.

16. OECD, *Prospects*, vol. 2, *Bulgaria*, pp. 150−63; RFER, *Bulgarian Situation Report*, 25, 8 September, 1976; 16, 3 March, 1977; 6, 6 March, 1983. For a comparison of Bulgaria's agricultural inputs and efficiency with the rest of Eastern Europe, including the USSR, see the United Nations Economic Commission for Europe, *Economic Survey of Europe, 1981* (New York: UN Secretariat, 1982), pp. 178−94; and *1983* (New York: UN Secretariat, 1984), pp. 182−90.

17. The same source finds that only a 44 per cent increase occurred between 1965 and 1975. See Thad Alton *et al.*, *Agricultural Output, Expenses and Depreciation, Gross Product and Net Product in Eastern Europe, 1965, 1970, and 1975−1981, Research Project on National Income in East Central Europe*, OP-71 (New York: L.W. International Financial Research, 1982), table 1, p. 16. For a Bulgarian accounting, see Todorov *et al.*, *Stopanska istoriia na Bulgariia*, pp. 566−71.

18. Jackson, 'Recent Economic Performance and Policy in Bulgaria', p. 22; *Statisticheski spravochnik na NR Bulgariia, 1984* [Statistical Handbook of the PR Bulgaria, 1984] (Sofia, 1984), table 10, p. 119.

19. OECD, *Prospects*, vol. 2, *Bulgaria*, pp. 179−83; *Statisticheski spravochnik, 1984*, table 26, p. 133.

20. OECD, *Prospects*, vol. 2, *Bulgaria*, pp. 165−72.

21. The following account is largely from ibid., pp. 175−6.

22. *Statisticheski spravochnik, 1984*, pp. 121−41; RFER, *Bulgarian Situation Report*, 6, 3 March, 1977.

8 FOREIGN TRADE AND DOMESTIC LIVING STANDARDS

The post-war pattern of Bulgarian production and investment described in Chapters 6 and 7 does not display striking differences, except for slightly higher rates of growth, from the Soviet experience. Western observers sometimes cite this similarity to support the notion that part of the Bulgarian government's unfailing loyalty to Soviet foreign policy derives from an economy that is not only tied closely to that of the USSR, but is also a mirror image of it. This is one of two chapters devoted to aspects of the Bulgarian economy that afford greater contrast with the Soviet pattern. This chapter examines the larger role of foreign trade and the steadier advance of living standards. Chapter 9 treats Bulgaria's more continuous attention to reforming the system. Both distinctions help to explain a better Bulgarian performance. But whether these distinctions explain more than do Bulgaria's smaller size, lower initial level of industrial development, and lower continuing level of military expenditure, compared with the USSR, we cannot say with certainty.

Three features of Bulgaria's foreign trade in the post-1960 period stand out. First is the well-known record of greater dependence on Eastern Europe in general, and on the Soviet Union in particular, than other members of the Council for Mutual Economic Assistance (CMEA). Secondly, Bulgarian exports have consisted of more agricultural goods, processed and unprocessed combined, than have those from any of the countries of eastern or southern Europe. This second familiar feature has facilitated a significant advance in the Bulgarian standard of living. Here, along with Hungary and its equally small population, is the only Eastern European economy to have fed its own population without recurring shortages or large-scale imports for the past two decades. An improved supply and variety of foodstuffs lies at the heart of the limited, but significant improvement in the country's overall standard of living that began during the 1960s. That improvement has become more noticeable during the past decade.

Less well known is a third distinction, the large relative size of Bulgaria's foreign trade. As may be seen in Table 8.3, the sum of exports and imports, or trade turnover, now matches the value of the net material product (NMP). This proportion places Bulgaria ahead of all other Eastern European economies. It also matches the ratio of turnover to gross domestic product, admittedly a larger aggregate than NMP, for all smaller Western European economies except Belgium, if we overlook artificial Eastern European exchange rates and also the Eastern European disparity between foreign trade and domestic prices. Bulgaria's present position is the result of rapid growth in trade turnover since the mid-1950s. Turnover growth averaged 16 per cent a year for 1955–70 and, as noted in Table 7.3, 10 per cent for 1971–80. Comparable figures for Eastern Europe as a whole were 9 per cent and 7 per cent, respectively, about the same as for Western Europe.

Foreign trade was of course growing faster than national income in all of these countries, suggesting a leading role for trade in economic growth across the continent. The elasticity of Bulgarian exports and imports with respect to national product (that is, the ratio of trade to NMP or GDP growth) was, however, consistently ahead of the European average. Bulgarian export elasticity stood at 1.7 for 1971–80, compared with an Eastern and Western European average of 1.4.[1] This rate of growth brought export value from 15 per cent of NMP in 1960 to 23 per cent by 1970, to 43 per cent by 1980 and to 50 per cent by 1983. The exact relationship of foreign trade to domestic development remains hard to quantify, but such a large surge in exports, half again the growth rate for NMP recorded in Table 7.3, must have played a positive part. This is true even when we reduce the above percentages to compensate for the Eastern European distortion of separate, often higher prices set for exports and imports as opposed to the prices fixed for the same goods domestically produced and consumed. The actual Bulgarian trade/income ratio since 1960 has probably still exceeded the 'expected ratio' calculated by the United Nations Economic Commission for Europe from population size and per-capita income.[2]

Bulgaria's rapidly growing foreign trade has affected domestic growth not just because of its size, but also because of recurring imbalances between exports and imports to different countries and regions. We need to know the changing structure of trade in order to appraise its effect on domestic growth and living standards — in other words, on investment and consumption. After all, the ratio of exports to roughly estimated national income had reached nearly 15 per cent in the last years before the First World War, as noted in

Chapter 1. Its rapid rise in those pre-war decades did facilitate the rapid monetisation of the Bulgarian economy, but hardly its industrialisation. How the present advance in the export ratio, well past the pre-1914 level, is connected to industrial investment and urban consumption helps to define the difference between the two periods. So does the new set of economic relationships with both Eastern and Western Europe that have emerged since the Second World War.

Dynamics of the Trade Balance

Bulgaria's post-war emphasis on foreign trade emerged, as might be expected in an economy based on the first principles of Soviet-style planning, from initial attention to imports rather than exports. Imports demanded as inputs for rapidly expanding industrial output were the first foreign-trade data entered into the projections for central planning. Unlike the Soviet Union, however, the Bulgarian list of necessary industrial imports included not only machinery, but also large quantities of fuel and other raw materials or semi-finished goods. The Bulgarian effort to manufacture import substitutes during the 1920s, it may be recalled from Chapter 2, had already prompted an equivalent upsurge in imports of industrial inputs. Many of the smaller developing economies in the Third World have faced a similar dilemma, to which more manufactured exports still seem their best answer.

Bulgarian demand for imported industrial inputs rose sharply during the 1960s and 1970s. First came the machinery, iron ore and even coal, needed for the growing production of ferrous metals. Machinery and equipment, primarily for industry, has accounted for almost half of total import value, as noted in Table 8.1. In absolute terms, it rose almost threefold during the 1960s and again during the 1970s. Imports as a share of industrial consumption of inputs climbed from 20 per cent to 28 per cent between 1955–7 and 1968–70, still short of the 40 per cent share of the 1920s.[3]

Energy supplies, primarily petroleum, have accounted for the largest proportional increase within the structure of Bulgarian imports. During the 1970s the energy share virtually doubled, from 13 per cent to 25 per cent. As a result, combined purchase of machinery and energy supplies took the largest share of imports, some two-thirds, of all the Eastern European countries. This

Table 8.1: Structure of Major Imports, 1955–83 (percentages)

	1955–7	1968–70	1975–7	1981–3
Machinery	47.7	50.7	49.1	47.3
Fuels, metals and minerals	29.9	27.2	25.4	25.3
Chemicals	9.9	9.0	6.1	7.1
Construction and non-food inputs	4.2	4.1	7.4	7.0
Foodstuffs	1.0	4.4	2.5	3.0
Processed foods			2.2	1.6
Consumer manufactures	6.5	4.0	6.2	7.8

Sources: United Nations Economic Commission for Europe (UNECE), *Economic Survey of Europe, 1971* (New York: UN Secretariat), vol. 1, p. 55; *Statisticheski godishnik na NR Bulgariia, 1982* (Sofia, 1983), p. 380; *Statisticheski spravochnik na NR Bulgariia, 1984* (Sofia, 1984), pp. 65–6.

Table 8.2: Structure of Major Exports, 1955–83 (percentages)

	1955–7	1968–70	1975–7	1981–3
Machinery	8.2	35.5	47.2	53.8
Fuels, metals and minerals	27.9	27.2	8.1	11.0
Chemicals	2.9	4.6	3.5	2.9
Construction and non-food inputs	8.0	4.7	4.1	3.0
Foodstuffs	28.0	35.1	4.8	3.1
Processed foods			20.8	13.6
Consumer manufactures	24.8	10.5	10.4	9.9

Sources: UNECE, *Economic Survey of Europe, 1971*, vol. 1, p. 55; *Statisticheski godishnik na NR Bulgariia, 1982* (Sofia, 1983), p. 380; *Statisticheski spravochnik na NR Bulgariia, 1984* (Sofia, 1984), pp. 65–6.

ranking follows logically from a degree of dependence on imported energy supplies, about 75 per cent of domestic consumption, which is also unmatched among the other Eastern European economies, typically well endowed with coal, if not much with oil. Only neighbouring Greece depends on imported energy to a similar extent. Bulgaria's efforts to reduce its dependence by the introduction of nuclear power have gone much more slowly than expected, still producing less than 10 per cent of the country's electricity.[4]

The exports generated to pay for these investment-oriented imports (barely 10 per cent have been consumer goods) have increasingly come from the sale of machinery, especially electronics equipment. According to Table 8.2, machinery's share of export value climbed spectacularly from 8 per cent for 1955–7 to 36 per cent in 1968–70 to 54 per cent by 1981–3. By itself, this relative growth suggests a continuing Bulgarian transition away from the agricultural exports upon which the country had depended before the Second World War to modern manufactures with more value added. Such a suggestion is in part deceptive. The role of agricultural exports, albeit over 80 per cent processed, was until 1980 the largest of any Eastern European country. Their 35 per cent share in 1968–70 had admittedly been cut to 26 per cent by 1975–7 and to 17 per cent by 1981–3.

These shrinking percentages understate the share of food exports in two ways, one potential and the other actual. First, they are based on sales which, like the rest of Bulgaria's export value, are 80 per cent to other Eastern European countries. These countries paid prices which were, in 1970, according to the calculations of an American economist, 26 per cent under world prices for comparable foodstuffs. But for all exports, Eastern European prices were 23 per cent *above* the world level. If revalued at world prices, Bulgarian agricultural exports would therefore have accounted for 43 per cent rather than 35 per cent of export earnings from trade with Eastern countries.[5] In fact, this extra amount was earning forgone by the Bulgarian export sector in its Eastern European trade and an incentive to increase sales to Western countries paying the world price. Bulgaria's terms of trade with Eastern Europe have led the decline for total trade noted in Table 8.3, slipping to 72 per cent of the 1970 level by 1980 and to 59 per cent by 1982. This drop confirms the relative lack of improvement in the price of agricultural compared with industrial or energy exports. The former was a relatively larger share of Bulgarian exports, and the latter a relatively larger share of its imports, than was the case for the rest of Eastern Europe.[6] Since 1975 the implicit price index for imports has risen 74 per cent, mainly because of Soviet oil, whereas the export index climbed only 9 per cent.

The second feature of Bulgaria's relatively large food exports to Eastern Europe has — until the early 1980s at least — compensated for losses attributable to prices below the world level. The scarcity of foodstuffs in much of Eastern Europe has made them 'hard goods' and allowed them to command non-price premiums. The premiums have taken a form described by Western economists as 'piggy-backing', that is, higher-priced, but lower-quality 'soft' machinery added onto the annual bilateral agreements with Bulgaria's Eastern European partners, in return for the delivery of eggs, meat, vegetables, cigarettes and other desirables.[7]

Since 1978 the country's Eastern European trade has run a

Table 8.3: Foreign Trade Balance and Terms of Trade, 1961–83 (in million leva)

A. Foreign trade balance

Annual average	1961–3 X–M	1968–70 X–M	1975–7 X–M	1981–3 X–M
USSR	457–506	1151–1121	2849–3024	5669–6172
Other CMEA[a]	257–251	448– 432	1246–1163	2010–2310[c]
Western	89–105	309– 366	513–1061	1267–1812
Third World	22– 16	99– 105	532– 243	1716– 618
Other[b]	60– 52	104– 67	114– 87	191– 24[c]
Totals	885–930	2111–2091	5254–5578	10853–10927

B. Terms of trade

	1960	1965	1970	1975	1980	1982
Terms of total trade[d]	102	104	100	96	76	65
Total X+M as % of NMP[e]	31	43	46	69	82	96

X–M = exports minus imports.
[a] Including Cuba.
[b] Mainly Yugoslavia.
[c] Approximate values.
[d] Export divided by import price index, 1970 = 100.
[e] Net material product.

Sources: *Statisticheski godishnik na NR Bulgariia, 1964*, p. 305; *1971*, pp. 195–7; *1982*, 382–4; *Vunshna turgoviia na NR Bulgariia, 1939–1975* (Sofia, 1976), pp. 26–8; *Statisticheski spravochnik na NR Bulgariia, 1984*, pp. 67–9; M. Allen, 'The Bulgarian Economy in the 1970s', in JEC, *Eastern European Economies Post-Helsinki* (Washington, DC: US Government Printing Office, 1977), p. 689; M.R. Jackson, 'Recent Economic Performance and Policy in Bulgaria', in JEC, *East European Economies: Slow Growth in the 1980s* (Washington, DC: US Government Printing Office, 1985), p. 13.

significant import surplus. To some extent this surplus is another indication of the economic value placed on its agricultural exports; it is also a reflection of political capital earned by loyalty to the Soviet Union. Accompanying this trade deficit with the Eastern European countries has been a small export surplus with Western Europe and a large one with the Third World. A sizeable deficit in Western European trade had appeared in the mid-1970s. The breakdown of Bulgarian trade balances in Table 8.3 makes it clear that significant regional deficits and surpluses have been typical of the period since 1970. The following analysis of economic relations with the three principal regions concentrates on this more recent period. The balance of payments, and not just the balance of trade, also become relevant; import surpluses require covering credits and eventual obligations of debt repayment.

CMEA Integration and the Soviet Connection

Bulgarian economic relations with the Soviet Union and the rest of Eastern Europe have become more complex with the trade imbalances of the past 15 years. Adding to this complexity since the late 1950s have been the efforts of the Council of Mutual Economic Assistance (CMEA) to integrate the Eastern European economies according to what Marxist terminology calls the international division of labour and what Western terminology calls comparative advantage.[8] Such specialisation has not gone far enough to eliminate the annual bilateral agreements by which each member determines its trade with every other, or to introduce a single, fully convertible currency in which surpluses in trade with one member can regularly be used to cover deficits with another. Thus the organisation has not made much progress toward the 'liberalisation of trade', which has been a hallmark of the Western European experience under the European Economic Community (EEC). The quotas implicit in these bilateral agreements continue to play the part of pre-war tariff barriers in restricting trade outside the agreements.

On the other hand, the purpose of CMEA since its emergence as an active organisation in the late 1950s has never been to create a Western-style customs union or to lay the groundwork for a socialist market economy operating according to world price signals. Instead, its emphasis has remained on the joint orchestration of the national planning mechanisms to select a few priorities for specialisation among the membership, and to eliminate some of the duplication fostered by the original Soviet model of balanced industrialisation, based on each country producing a full range of all major goods. The movement of capital and labour between members to collaborate on joint investment, mainly Soviet projects for raw materials, has remained far smaller than in the market economies of the EEC. Specialisation has not gone so far as to eliminate major areas of unprofitable industrial production in any member country (for example, Bulgarian ferrous metallurgy). It had gone far enough, by 1983, however, to force the Soviet Union to abandon plans for replacing imports of Bulgarian fork-lift trucks with its own production from a projected new plant at Kiev.

CMEA's existence has all the same encouraged a rate of trade growth roughly comparable to the Western European one. This alone, in the judgement of one Western economic historian, has set the post-war period far apart from the sad Eastern European experience with economic nationalism and reduced trade levels during the inter-war period.[9]

Bulgaria's entire post-war experience has of course been characterised by a closer economic relationship with the USSR, beyond trade alone, than has any other Eastern European country's. Regardless of the Soviet political motives stressed by Western observers or the joint CMEA membership stressed by the Bulgarian side, economic relations between the largest and smallest of the Eastern European economies — one rich in industrial raw materials and the other rich in agricultural resources — should rationally have been large. Political alliance nevertheless encouraged trade and seems the best explanation for the movement of capital and labour that has occurred. After taking virtually no reparations from post-war Bulgaria, in contrast to Romania, the Soviet Union afforded Bulgaria's heavy industry credits worth 2,200 million roubles (in 1950 rates) between 1947 and 1957. Several electric power plants, the Varna shipyards, agricultural machinery in Ruse and mining equipment in Sofia, plus the Lenin metallurgical complex near there, received the bulk of these credits. The five joint Soviet-Bulgarian companies made little positive contribution (see Chapter 6), but at least were not exploitative as were the Sovroms in Romania.

During the period 1958–75, more clearly within the CMEA framework, the Soviet Union dispersed another 2,200 million roubles worth of credits. New joint projects with the USSR were two textile plants, production of the Balkancar electric fork-lift trucks in Sofia, more machine manufacturing in Ruse, and the Kremikovtsi metallurgical complex. In the most recent decade, major new Soviet projects have been the three proposed nuclear power plants (one is now operating) and the new metallurgical complex at Burgas (not yet open and now relying on East German machinery). All credits have carried nominal interest rates of 2–3 per cent, but have been due for repayment in just ten or twelve years. Although these form a small share of Soviet resources, the contribution of these credits to Bulgarian industry cannot be dismissed lightly. They reportedly amounted to 27 per cent of total investment for 1948–56 and 15 per cent for 1958–75, including one-third of the value of new machinery and equipment during the latter period.[10]

The above projects have brought an undetermined number of Soviet engineers and technicians to Bulgaria, sometimes for several years at a time, but rarely with public notice. Since the 1960s a reverse flow of skilled Bulgarian workers, mainly in construction, to the Soviet Union has swamped the Soviet totals. Bulgarians working on several CMEA projects in the USSR reached 10,000 by 1975; the official total was 27,000 by 1984.[11] The absence of these skilled workers, together with another 20,000 primarily in East Germany, Cuba and Libya (the only non-member of CMEA), has become a serious drain on a domestic labour force already stretched too thin (see Chapter 7).

The principal area of Bulgarian integration with the Soviet and other Eastern European economies has remained the planning of their joint trade for several Five-Year Plans in advance. Since 1956, this has been done through multilateral CMEA negotiations as well as through bilateral written agreements. The earliest evidence of a CMEA session seeking to synchronise Five-Year Plan trade and production for at least some branches of industry dates back to 1954 (see Chapter 6). By 1957, a session in Sofia approved the Bulgarian proposal to expand its CMEA exports of fruit and vegetables in return for Soviet metals and raw materials, Czech cellulose and paper, and East German chemicals and minerals. What has developed since then is a series of 'specialisation agreements' for each branch of production, which are incorporated into the annual bilateral trade pacts.[12] In electronics, for instance, some 121 agreements for producing specific items accounted for 52 per cent of the value of Bulgarian export earnings in 1979. The figure for machinery exports exceeded 40 per cent. By the 1980s, Bulgaria was responsible for 40 per cent of the export of both electric hoisting and data-processing equipment. Overall, according to one Bulgarian source, 30 per cent of trade turnover with CMEA partners and 35 per cent with the USSR were tied to specialisation agreements by 1980.[13]

With or without these agreements, Bulgarian trade with the USSR has proved to be so large as to raise speculation among Western economists about the integration of the former's economy into the latter's. Certainly the predominant position of the Soviet

Union in CMEA and in bilateral relations has had a major impact on the shape of Bulgarian foreign trade and also transport. By 1975, trade with the USSR accounted for 77 per cent of Bulgaria's transport tonnage, mainly as a result of Soviet imports. Some 21 bilateral agreements provided for this traffic. Its movement was mainly across the Black Sea, and even before the opening of the ferry service between Varna and Iliechovsk, near Odessa, in 1978, it was responsible for increasing Bulgarian imports brought in by ship to nearly 90 per cent of total tonnage.[14] The railway share fell accordingly. But the leap from these ties to full-scale integration, where each country abandons its existing lines of less efficient production, has simply not occurred.

Throughout the 1970s, the USSR none the less continued to purchase half of Bulgaria's machinery exports and one-quarter of its electronics exports within CMEA. Food products accounted for only 23 per cent of Bulgarian export value within CMEA over the same period, less than either of the first two branches, despite the latter's more complementary role as scarce, or 'hard' goods for the USSR. Soviet purchases accounted for most of the overall Bulgarian exports of cigarettes, eggs, canned meat and vegetables. Other food exports have gone much further afield. For the 1970–80 period, 77 per cent of their value was sold outside the CMEA, primarily, as we shall see, to Western Europe. Only Bulgarian exports of non-food consumer goods fit a pattern of integration with the USSR. Two-thirds of Bulgaria's overall exports in this category are sent there. Much of their production is based on inputs of cotton, leather, wood or other raw materials that are imported from the USSR specifically for processing on the *ishleme* basis first practised in the Ottoman period (see Chapter 1).[15] Yet light industry other than food comprises just 10 per cent of total Bulgarian production.

The bulk of Soviet imports into Bulgaria have consisted instead of industrial raw materials, primarily iron ore and petroleum. These imports have above all supported the continuing Bulgarian effort to develop its own production of ferrous metallurgy, despite the absence of comparative advantage. Soviet prices for petroleum, which were set substantially below world levels until the late 1970s and somewhat below since then, made the agreed quotas for each Five-Year Plan a subsidised encouragement to Bulgarian heavy industry.[16] Since 1976, any additional oil sales have been made at world prices, and since 1981 agreed quotas have been held constant. Thus the planned growth of Bulgarian industry, and producers' goods in particular, can no longer proceed on the assumption of comparable increases in Soviet oil supplies. Nor are Soviet negotiators reportedly now willing to accept inferior sorts of Bulgarian machinery in return for these supplies.

Other problems have also plagued Bulgarian foreign trade within the CMEA framework. Exported manufactures have been kept too isolated from world prices and their assortment too wide for an increasingly specialised world market. Bulgarian and Western economists agree that this isolation and 'universalism' have inhibited the matching of Western best practice in machinery or electronics production.[17] Too many prices and products have remained unchanged for too long under a CMEA framework geared to Five-Year Plans. The importance of Bulgaria's relatively small Western trade lies precisely in the access to more recent technology and to a more flexible supply of imported inputs for industry.

Economic Relations with the West and the Third World

The first significant change in post-war Bulgarian trade outside Eastern Europe did not occur until the late 1960s. The increase in Western imports, primarily from West Germany, was responsible. The Western share of 14 per cent of Bulgarian imports in 1960 had grown to 19 per cent by 1970. It peaked at 24 per cent in 1975. As recorded in Table 8.4, Bulgarian exports to Western markets did not increase proportionally over the same period. By 1975 they had slipped to 9 per cent of total export value. Since then the deficit in Western trade was briefly replaced by a surplus, but a small negative balance has reappeared since 1980. In the absence of a convertible currency, Bulgaria could only cover these trade deficits with its Western partners by borrowing the money from them. This section must therefore treat the rise and at least temporary resolution of a Bulgarian problem of debt repayment, hardly as serious as the famous Polish case, but still significant. Bulgarian exports to Third World countries grew from only 3 per cent of the total in 1960 to 6 per cent in 1970 to 11 per cent in 1975 and 13 per cent for 1980–3, mainly because of increased sales and projects in Libya and Iraq. These exports have done little, however, to relieve the problem of paying for Western imports. The large export surplus that has resulted in trade with the Third World has not in the main

Table 8.4: Direction of Foreign Trade, 1960–83 (in percentages)

Imports	1960	1965	1970	1975	1980	1983
USSR	52.6	50.0	52.2	50.7	57.3	58.5
GDR	11.1	7.2	8.6	6.3	6.6	5.7
Other CMEA[a]	21.6	17.1	15.4	15.0	15.0	16.1
FRG	5.9	5.8	2.7	7.8	4.8	3.9
Other Western[b]	7.7	16.5	16.5	15.8	12.4	9.8
Libya	—	—	—	—	1.0	2.0
Other Third World[c]	2.4	3.5	4.7	3.9	2.9	4.1
Exports						
USSR	53.8	52.1	53.8	54.6	49.9	58.5
GDR	10.2	9.2	8.7	7.0	5.5	5.7
Other CMEA[a]	20.0	18.0	16.8	18.4	18.0	14.9
FRG	3.3	3.5	2.6	1.7	2.5	1.7
Other Western[b]	9.1	12.3	11.6	7.6	13.3	8.8
Libya	—	—	—	2.2	3.6	4.1
Other Third World[c]	3.4	4.6	6.1	8.5	9.8	9.0

[a] Including Cuba.
[b] Mainly Switzerland and Italy.
[c] Mainly Iraq, Iran, Turkey and Algeria.

Sources: *Statisticheski godishnik na NR Bulgariia, 1964*, p. 305; *1971*, pp. 195–7; *1982*, pp. 382–4; *Vunshna turgoviia na NR Bulgariia, 1939–1975* (Sofia, 1976), pp. 26–8; *Statisticheski spravochnik na NR Bulgariia, 1984*, pp. 67–9; Allen, 'The Bulgarian Economy in the 1970s', p. 689; Jackson, 'Recent Economic Performance and Policy in Bulgaria', p. 13.

been available to Bulgaria in convertible currency.

The rise in Western imports during the period 1965–75 derived mainly from Bulgaria's aforementioned demand for a flexible supply of inputs to support industrial development. Capital goods, primarily sophisticated machinery, accounted for 30–50 per cent of Western imports, chemicals another 10–30 per cent, and special metals 10–15 per cent. Consumer goods were typically a small and residual fraction. West Germany furnished 30–40 per cent of the Western total. Italy, France and Switzerland provided another quarter. To cover the purchases, Bulgaria relied on foodstuffs for 40 per cent of its exports to West Germany, and only 20 per cent to the other three main partners. Textiles typically covered another 20 per cent of hard-currency exports, with metals and machinery about 10 per cent a piece.[18] Deficiencies in punctual delivery, packaging and advertising contributed as much as the standard explanations of inferior Bulgarian quality or EEC trade restrictions to the failure of Bulgarian exports to match Western inputs over this decade. In 1970, an infusion of Soviet hard-currency credit allowed Bulgaria to pay off the Western debt accumulated during 1966–8. In return, Bulgaria ran a large export surplus in its trade with the USSR from 1970 to 1973.

Since 1975, West German imports have slipped to one-quarter of the Western aggregate. Austria, Japan and the United States have stepped in to take up about one-fifth of that total. American imports, however, have consisted mainly of feed grains, rather than industrial inputs. American trade has been hampered not only by Bulgarian deficiencies in exporting competitive products, but also by official US denial of most-favoured-nation (MFN) status and the preferential tariffs that go with it.[19]

The reduction of Bulgaria's Western trade deficit and the elimination of its overall deficit since 1975 still left substantial short-term debts owed to Western European banks from borrowing over the previous decade. By 1976, the Bulgarian debt had climbed to 13 per cent of estimated gross national product and the annual debt-service ratio to 44 per cent of hard currency, that is, Western export earnings. The former ratio was the highest in Eastern Europe at the time. The latter was triple the inter-war peak in the Bulgarian debt-service ratio to all export earnings in 1931–2 (see Chapter 3). By 1981, however, a Bulgarian debt of $3,000 million had been cut in half. According to American analysis, this reduction was largely the result of extra Soviet deliveries of petroleum, which were in turn sold as refined products for hard currency to Western buyers.[20] These deliveries helped to create a new Bulgarian trade deficit with the USSR, however. The recent Soviet desire to repair this deficit and the growing limits on Soviet petroleum production since 1980 make it unlikely that such largesse can be counted on another time.

Two major Bulgarian strategies to avoid Western credit and still maintain access to Western imports and technology have been the promotion of tourism and joint-investment ventures. Extensive hotel, restaurant and road construction on the Black Sea coast helped to double the number of Western tourists between 1960 and 1970. Since that time, the pace of construction for tourism has accelerated. However, other Eastern European countries, plus neighbouring Yugoslavia, Greece and especially Turkey, have been solely responsible for doubling the annual number of visitors between 1970 and 1980.[21] In the absence of published Bulgarian data on the balance of payments, we can only speculate on how much hard currency Western tourists have contributed to current-account earnings over time.

The other Bulgarian strategy for covering the cost of hard-currency imports has been to promote direct Western investment under licence or through joint ventures. Legal provisions for both were introduced during the mid-1960s. An initial joint venture with the French car manufacturer Renault failed within a few years, but a number of licensing agreements soon worked out to the satisfaction of both Bulgarian and Western enterprises. Agreement with Coca-Cola has been followed in more recent years by similar licences for Schweppes and Pepsi-Cola. By the late 1970s, licensing agreements had spread into industrial manufacture, including the fork-lift truck firm, Balkancar. Some 40 joint ventures were operating in trade and tourism, including hotel construction.

Direct Western investment in joint industrial ventures remained miniscule, perhaps prompting the new Bulgarian legislation of 1980. Its terms allow foreign partners to claim majority rights in controlling the enterprise and in receiving hard-currency earnings. Bulgarian officials had initially hoped to attract Western capital and technology for raw-material extraction or processing for export to third countries. The four ventures signed to date, including three with Japanese partners, have dealt instead with the manufacture of electronics, plastics and consumer goods.[22] Although 20 joint trading ventures have now been concluded with West Germany, Italy and France, Bulgarian exports to these countries have not increased since 1970. Significant import substitution or a reduction in export of domestically scarce goods like cement or fertiliser must await, respectively, a higher level of Western investment or a lower level of Eastern European demand than has so far been the case.

Living Standards and Consumption Patterns

Bulgaria'a post-war standard of living has been slow to benefit from the country's booming foreign trade, at least on the import side. The share of consumer goods in total imports stayed too small, about 10 per cent, to add significantly to domestic consumption or to stimulate import substitution along the lines of the Yugoslav experience. On the export side, however, a real connection appeared by the 1960s. Large exports of foodstuffs raised rural incomes, and rising production left a surplus to improve the urban supply of food. Since 1970, the relationship of exports to living standards has been more complex.

No significant upturn in Bulgaria's post-war standard of living, official figures not withstanding, was achieved until the 1960s. Standards had probably fallen from 1948 to 1952, largely because of agricultural problems during collectivisation. Collective-farm income rose only 35 per cent from its low 1951 level by 1958. Mitigating a 10 per cent annual rise in largely industrial wages over this period were hidden price increases and the continuing scarcity and poor quality of most consumer goods.[23]

Improvement began after 1960, first with agricultural income. Its real increase over the period 1960–70 averaged 6.7 per cent a year, compared with 4.9 per cent for real industrial wages. Average peasant income rose from one-third to two-thirds of wages for the urban factory worker in money terms for 1952–68, and came even closer, if self-consumption of agricultural produce could be fully calculated. By the mid-1960s, Western visitors to the Bulgarian countryside, including the author, were struck by the number of new houses under construction and the number of new cars parked in front of them. The improvement reckoned by a Bulgarian economic historian in the rural standard of living between 1939 and 1968, which consisted of a doubling of real income and a 33 per cent rise in food consumption, occurred largely during the last decade of the period.[24] By 1970, a corresponding improvement in urban food supplies made the decade's increase in industrial wages meaningful. The share of foodstuffs in personal consumption fell from 52.5 per cent to 48.3 per cent for 1960–8, and then to 44.5 per cent by 1977; meanwhile the average daily consumption of calories had climbed to a Western European level of 3,500.[25]

During the period 1970–83, urban workers saw their money wages increase by 3.9 per cent a year, reaching a monthly average of 197 leva, but real wages rose only by a 2 per cent annual average. Responsible for this return to the modest increments of the early 1960s was the tiny 0.5 per cent rise in real wages for 1976–80 recorded in Table 7.3, the lowest annual rate in Europe. The cause now was not massive new metallurgical investment, as in 1962–3 (see Chapter 7), but rather a price increase of at least 25 per cent for foodstuffs and of 15 per cent for all consumer goods. These price rises were imposed to facilitate foreign trade and to generate a trade surplus by bringing Bulgarian prices more into line with both world levels and the domestic cost of production. Data from the United Nation's International Labour Organization find major food prices for 1981 consistent with French levels, despite the fact that wages for Bulgarian manufacturing remain about one-quarter of the French level.[26]

Given these limitations, it is unlikely that the large share of Bulgarian personal consumption devoted to foodstuffs has declined further since 1977. That was a level similar to others in Eastern Europe, but over half again the Western European average of about 25 per cent. None the less, progress has been made in the composition of the Bulgarian diet. According to Table 8.5, the cereal share has declined in favour of meat, but not enough. Meat consumption per capita rose by 70 per cent for 1970–83, despite a 3 per cent drop in 1980. Poultry was the chief gainer. The meat aggregate still covered only seven-eighths of the 'scientific norms' set by the United Nations. Milk and egg consumption also rose sharply, but fell short of UN norms. Fruit and vegetables barely met half of these norms, but at least ranked well ahead of the levels elsewhere in Eastern Europe.[27]

Deficiencies may be traced not only to export diversion, but also to retail networks for domestic distribution. The number of foodstores increased by 25 per cent during the 1970s and their floor space increased even more, because of new supermarkets. Yet the typical shop remains small and poorly organised by Western standards. Separate lines to select, pay for and pick up purchases delay customers needlessly. Self-service still accounts for only one-third of retail sales. Bad packing and inconsistent shipment of produce, alternating shortages with unsaleable gluts, hit perishable fruit and vegetables the hardest.[28] Sofia and the larger towns had more supermarkets by the end of the decade, but also paid higher prices for what was available. All foodstuffs continued to be vulnerable to diversion to the tourist hotels on the Black Sea coast. Since the late 1970s, however, the relative regularity of Bulgarian food supplies has stood in sharp contrast to neighbouring Romania and has attracted customers from there and sometimes from Yugoslavia.[29]

Other consumer goods, more specifically durables like clothing and household appliances, did not record a real upturn until the 1970s. Then private cars — mainly Soviet Fiats, partly produced in Bulgaria — and refrigerators led the advances noted in Table 8.5. Car auto sales exceeded half a million cars, before dropping off abruptly to modest annual increases since 1978. Sofia far surpasses the national average of 34 cars per 100 households. Its 350,000 private vehicles by 1980 joined with public transport to create a very modern problem for the Bulgarian capital city. Leaded petrol has compounded air pollution, already a serious problem, caused by burning lignite for heat.

Until the past few years, the smaller durables like clothing, shoes and household appliances have ranked low in quality and variety. Both have improved since 1980 to an extent that has impressed Western visitors, again including the author. Prices for the better-quality items are also high, for instance, 50–75 leva for shoes or dresses, when monthly salaries still average only 200 leva. Yet these items do not go unsold, but are bought by people in households with several incomes and low housing costs. They are also sold for Bulgarian leva, rather than for hard currency in the Corecom shops, which have introduced the urban elite to Western consumer goods since the 1970s. Chapter 9 will deal with the policy decisions and managerial reorganisation that have generated such home-made goods. For the time being, we must categorise their appearance as 'import emulation', rather than substitution. They are taking the place of the larger fraction of consumer goods in imports (up to 30 per cent) that had been planned, but never purchased.

Housing for a largely urban population and facilities for a comprehensive set of social services have continued to be the most serious shortcomings in the Bulgarian standard of living. Their construction is classified, according to the Soviet pattern, under services, and therefore is treated as a residual in a planning and accounting process organised around producing goods. Schools and hospitals are overcrowded and often poorly equipped. Most elementary schools in Sofia and the larger towns still operate on split shifts. Residential housing has received ambitious targets in recent Five-Year Plans, but has typically failed to meet more than two-thirds of the planned apartment total. In 1975, Bulgaria's average of 22.5 units of housing per 1,000 population still ranked bottom in Eastern Europe, below even Romania's 26.5.[30] Since then apartment construction, typically for a form of condominium ownership based on fifteen-year mortgages for prices ten times annual income, has received higher planning priority. The current goal is one room for each family member by 1990. Targets are still not being met, with the shortfalls exceeding 20 per cent. Despite 650,000 new units built for two million people between 1972 and

Table 8.5: Indicators of Living Standard, 1956–83

A. Per capita consumption (in kilograms)

	1956	1970	1983	UN norm
Meat	27	41	70	80
Milk, dairy[a]	81	117	183	260
Eggs[b]	69	122	225	265
Fish	—	6	8	10
Vegetables	74	89	109	180
Fruit	84	119	115	200
Sugar	12	33	36	32
Bread	257	239	209	135
Cotton, wool Knitwear[b]	3	8.5	13.2	
Footwear[b]	0.6	1.7	2.3	

B. Consumption index (1968=100) (Alton)

	1960	1977
Food	78	121
Clothing	62	126
Kitchen durables	63	194
Transport	41	289
Housing	81	134

C. Consumer durables (per 100 households)

	1965	1970	1983
Radios	59	62	92
Televisions	8	42	87
Electric washing machines	23	50	81
Refrigerators	5	29	88
Private cars	2	6	34

D. Housing and services (per 1,000 population)

	1956	1970	1983
Urban housing[c]	5.8	9.7	14.3
Retail shops	2.9	3.8	4.5
Hospital beds	5.0	8.4	9.0
Physicians		1.9	4.5
High-school students	165	182	172
University students	4.8	10.5	9.6

Table 8.5: Indicators of Living Standard, 1956–83 — *continued*

E. Consumer price indices (1960 = 100)

	1965	1970	1975	1980
Official	107	110	112	137
Alton	110	125	142	161

[a] In litres.
[b] In units.
[c] In square metres per capita.

Sources: *Statisticheski spravochnik, 1976*, pp. 88, 105; *1984*, pp. 171–4, 191, 199, 212, 218–20; T. Alton *et al.*, *Personal Consumption, in Eastern Europe, 1960–1978*, OP-57 (New York, 1979), pp. 22–8; T. Alton *et al.*, *Official and Alternative Consumer Price Indices in Eastern Europe*, OP-73 (New York, 1982), p. 14.

1982, qualified households wait several years for apartments in large towns, and sometimes five or ten years in Sofia, where the overcrowding is most serious.

The capital's growth from 650,000 in 1960 to 1.3 million by 1980, as both an industrial and administrative centre, would have posed a housing problem under any political system. The continuing export of domestically scarce cement to the Soviet Union may be cited as one political component of the housing shortage. More important, however, has been the shortage of skilled labour, resulting basically from rapid industrialisation and slow population growth, as noted in Chapter 7. The diversion of some skilled construction labour abroad for projects undertaken for reasons of foreign policy, or at home for prestige projects like the new National Centre of Culture (NDK) complex in Sofia, have, on the other hand, aggravated the labour shortage for residential construction.

One approach to the general labour shortage has been to cut back on the number of university students by reducing the often-abused time-limit for completing degrees and by making the entrance requirements harder. Similar restrictions have appeared across Europe, from France to the USSR, for varying reasons. Enrolment at the main university in Sofia has fallen by 25 per cent since 1978, the peak year for new, often overqualified and over-age graduates. In the early 1980s, an educational reform was launched in the high schools to make the final year mainly vocational training in a skilled trade. By this time, however, many of the urban elite, like their counterparts elsewhere in Europe, have come to regard a university education as something to which their children are entitled. Broader access to a university education, after all, has been the major change, together with greater stress on science and technology, made in the country's educational system during the Communist era.[31] Educational reform has not therefore been widely welcomed, and the consequences of these initial changes remain unclear.

The availability of housing and the price of food aside, the Bulgarian standard of living seems to have risen again, after the restrictive period of the late 1970s. The minimal increase in real wages for 1976–80 was accompanied by declining consumption in several important areas, culminating in the meat shortage of 1980. In the process, however, the Bulgarian economy was able to use this reduced consumption to trim the pressure on its balance of payments. We now turn to the recurring set of reforms by which Bulgarian authorities hope to achieve both rising domestic consumption and the competitive exports needed to protect the balance of payments. If economic reform is to succeed, however, the system of higher education will have to pay as much attention to training better managers as it now does to enlarging the labour force.

Notes

1. Nita Watts, 'Eastern and Western Europe', in Andrea Boltho (ed.), *The European Economy: Growth and Crisis* (Oxford: Oxford University Press, 1982), pp. 278–80; United Nations Economic Commission for Europe *Economic Survey of Europe, 1981* (Geneva: United Nations, 1982), p. 282.

2. The actual ratios during 1963–7 for East Germany, Czechoslovakia, Hungary and also for Spain, Portugal and Greece were still less than the calculated ones, according to the United Nations Economic Commission for Europe (UNECE), *Economic Bulletin for Europe*, vol. 21, no. 1 (Geneva: United Nations, 1970). By then the Bulgarian ratio had risen to 1.13. A useful discussion of these ratios is Edward A. Hewett, 'Foreign Trade Outcomes in Eastern and Western Economies', in Paul Marer and John Michael Montias (eds.), *East European Integration and East-West Trade* (Bloomington, Indiana: Indiana University Press, 1980), pp. 41–68.

3. UNECE, *Economic Survey of Europe, 1971* (New York: United Nations, 1972), p. 57.

4. Bulgarian expectations for deriving half of the country's electricity from nuclear power have now been put off from 1990 to 2000. On energy imports, see Marvin R. Jackson, 'Bulgaria's Economy in the 1970s: Adjusting Productivity to Structure', in Joint Economic Committee of the US Congress (JEC), *East European Economic Assessment*, pt 1 (Washington, DC: US Government Printing Office, 1981), p. 602; Nikolai Todorov et al., *Stopanska istoriia na Bulgariia, 681–1981* [Economic History of Bulgaria, 681–1981] (Sofia, 1981), pp. 526, 588–9.

5. See the discussion by Jan Vanous, in Marer and Montias, *East European Integration*, p. 315.
6. The Bulgarian terms of trade held virtually steady for the 1960s, falling by only 2 per cent. Jackson, 'Bulgaria's Economy in the 1970s', p. 597; and Mark Allen, 'The Bulgarian Economy in the 1970s', in Joint Economic Committee of the US Congress, *East European Economies Post-Helsinki* (Washington, DC: US Government Printing Office, 1977), p. 689.
7. Gur Ofer, 'Growth Strategy, Specialization in Agriculture, and Trade: Bulgaria and Eastern Europe', in Mater and Montias, *East European Integration*, pp. 299–313.
8. Cuba, Mongolia and Vietnam are CMEA members, together with the Eastern European states of Poland, Czechoslovakia, East Germany, Hungary, Romania, Bulgaria and the USSR. For a Western survey of the organisation and its activities, see Jozef M. Van Brabant, *Socialist Economic Integration* (Cambridge: Cambridge University Press, 1980). A Bulgarian critique of Western analysis is Zhak Ario (ed.), *Ikonomicheskoto sublizhavane na stranite ot sotsialisticheskata obshtost* [Economic Convergence of the Countries of the Socialist Camp] (Sofia, 1983), pp. 218–35.
9. See Sidney Pollard, *Peaceful Conquest, The Industrialization of Europe, 1760–1970* (Oxford: Oxford University Press, 1981), pp. 309–34.
10. Stefan Sharenkov and Vasil Sirov, 'Participation of the PR Bulgaria in the CMEA', and also Zlatko Zlatev, 'Socialist Industrialization in Bulgaria', in Mito Isusov (ed.), *Problems of the Transition from Capitalism to Socialism in Bulgaria* (Sofia, 1975), pp. 353–60 and 177–85, respectively.
11. Allen, 'Bulgarian Economy in the 1970s', p. 665; *Zemedelsko zname*, 6 November 1984.
12. Sharenkov and Sirov, 'Participation of Bulgaria in the CMEA', pp. 345–9; Triatko Bozhkov, *Nauchno-tekhnicheska i proizvodna integratsiia mezhdu NR Bulgariia i SSSR* [Scientific-Technical and Production Integration between the PR Bulgaria and the USSR] (Sofia, 1976), pp. 174–217.
13. L. Radulov (ed.), *Sotsialno-ikonomicheskata politika na Bulgarskata durzhava, 681–1981* [Socio-economic Policy of the Bulgarian State, 681–1981] (Varna, 1981), pp. 497–8.
14. Sea freight to and from Bulgaria jumped 19-fold for the period 1958–70, while rail freight barely doubled. The latter increased only 10 per cent for 1970–8. Truck transport grew by 40 per cent, after a sevenfold leap for 1958–70, mainly because of exports and internal trade. Todorov *et al.*, *Stopanska istoriia na Bulgariia*, pp. 513–17, 526; Bozhkov, *Nauchno-tekhnicheska integratsiia*, pp. 182–4.
15. Bozhkov, *Nauchno-tekhnicheska integratsiia*, pp. 176–8.
16. See Michael Manese and Jan Vanous, 'Soviet Policy Options in Trade with Eastern Europe', in Joint Economic Committee of the US Congress, *The Soviet Economy in the 1980s: Problems and Prospects*, pt 1 (Washington, DC: US Government Printing Office, 1982), pp. 102–15.
17. Alec Nove, *The Soviet Economic System* (Oxford: Oxford University Press, 1977), pp. 178–81; Todorov *et al.*, *Stopanska istoriia na Bulgariia*, pp. 591–2.
18. George W. Feiwel, 'Economic Development and Planning in Bulgaria in the 1970s', in Alec Nove, Hans Hermann Höhemann, Gertrud Seidenstecher (eds.), *The East European Economies in the 1970s* (London: Butterworths, 1982), pp. 223–5.
19. Deborah A. Lamb, 'Bulgaria: Performance and Prospects in Trade with the United States and the West', in Joint Economic Committee of the US Congress, *East-West Trade: The Prospects to 1985* (Washington, DC: US Government Printing Office, 1982), pp. 37–9.
20. See Jackson, 'Bulgaria's Economy', pp. 593–603; and Chapter 3 above.
21. *Statisticheski goadishnik na NR Bulgaria* [Statistical Yearbook of the PR

Bulgaria] *1982* (Sofia, 1983), p. 461; Paul Marer, 'Tourism in Eastern Europe', in Joseph Brada (ed.), *Quantitative and Ananlytical Studies in East-West Economic Relations* (Bloomington, Indiana: Indiana University Press, 1976), pp. 37–116.

22. Michael C. Kaser, 'The Industrial Enterprise in Bulgaria', in Ian Jefferies (ed.), *The Industrial Enterprise in Eastern Europe* (New York: Praeger, 1981), pp. 91–2; Sofia Press Agency, 'Bulgaria's Presence in Five Continents', 111, 103 (1983); RFER, *Bulgarian Situation Report*, 5, 18 April 1980; and 12, 23 July 1982.

23. J.F. Brown, *Bulgaria under Communist Rule* (New York: Praeger, 1970), pp. 43–6.

24. Liuben Berov, 'Realnite dokhodi i potreblenieto na trudeshite i seliani v Bulgariia prez 1938–1969' [Real Income and the Consumption of Workers and Peasants in Bulgaria, 1938–1969], *Problemi na truda*, vol. 1 (1970), pp. 23–34. Also see Todorov et al., *Statisticheski godishnik na NR Bulgariia*, pp. 531–2; and George W. Feiwel, *Growth and Reforms in the Centrally Planned Economies: The Lessons of the Bulgarian Experience* (New York: Praeger, 1977), pp. 214–24.

25. Organization for Economic Co-operation and Development (OECD), *Prospects for Agricultural Production and Trade in Eastern Europe*, vol. 2, *Bulgaria* (Paris: OECD, 1982), pp. 145–8; Thad Alton et al., *Personal Consumption in Eastern Europe, 1960–1978*, OP-57 (New York: L.W. International Financial Research, 1979), p. 18.

26. *European Marketing Data and Statistics, 1983* (London: Euromonitor Publications, 1983), pp. 177, 181. Also see Feiwel, 'Economic Development in Bulgaria', pp. 227–30; *Statisticheski godishnik na NR Bulgariia, 1982*, p. 367; Thad Alton et al., *Official and Alternate Consumer Price Indices in Eastern Europe, Selected Years, 1960–1981*, OP-73 (New York: L.W. International Financial Research, 1982), p. 14.

27. OECD, *Prospects*, pp. 145–8; Todorov et al., *Statisticheski godishnik na NR Bulgariia*, pp. 600–1; RFER, *Bulgarian Situation Report*, 4, 25 March 1983; UNECE, *Economic Survey of Europe, 1983* (New York: United Nations, 1984), pp. 178–81.

28. Todorov et al., *Stopanska istoriia na Bulgariia*, pp. 578–98; RFER, *Bulgarian Situation Report*, 12, 10 September 1981.

29. Marvin R. Jackson, 'Recent Economic Performance and Policy in Bulgaria', Joint Economic Committee of the US Congress (JEC), *East European Economies: Slow Growth in the 1980s*, pt 3 (Washington, DC: US Government Printing Office, 1985), pp. 15–23; Marvin R. Jackson, 'Romania's Economy at the end of the 1970s: Turning the Corner on Intensive Development', JEC, *East European Economic Assessment*, pt 1 (Washington, DC: US Government Printing Office, 1981), pp. 231–98.

30. Feiwel, *Growth*, p. 221; and also his 'Economic Development in Bulgaria', p. 230; Todorov et al., *Stopanska istoriia na Bulgariia*, p. 598; RFER, *Bulgarian Situation Report*, 18, 31 October 1978; and 4, 25 March 1983. The average size of new dwellings in 1981 remained — together with that of the USSR — Europe's smallest: 2.3 rooms, excluding kitchens. *European Marketing Data, 1983*, p. 299.

31. The number of university-level students per 10,000 population rose from 16 in 1939 to 55 in 1948, 70 in 1960 and 122 by 1974, before decreasing, first because of demographic decline and then as part of educational reform, to the 1983 level of 94. Primary and secondary-school students per 10,000, on the other hand, have returned to the 1939 level of less than 1,700, after reaching 2,000 in 1964. *Statisticheski godishnik na NR Bulgariia, 1982*, pp. 22–3. Illiteracy had already been reduced by 1934 to 31.5 per cent of the population over 7 years of age, the lowest rate in any of the Balkan states. By 1965, the Bulgarian figure was just 8.3 per cent, again the lowest in the region. Paul Shoup, *East European and Soviet Data Handbook* (New York: Columbia Univeristy Press, 1981), p. 169.

9 ECONOMIC REFORMS SINCE 1960

The general phrase 'economic reform' has a specific meaning for Eastern Europe and the Soviet Union. Since the 1960s, it has been synonymous with the streamlining of state planning and the decentralisation of ministerial controls. Its goal has been to create greater efficiency, not a market economy. During the first years following the Second World War, all of Eastern Europe adopted the Soviet system of central planning. In Bulgaria, the system's capacity for promoting the extensive growth of a few branches of industry, based on massive infusions of labour and capital, was well demonstrated (see Chapter 6). By the early 1960s, however, it was equally clear that sustained growth in all branches of production, including agriculture, would require changes in the original system. The looming shortage of labour and the growing importance of foreign trade (detailed in Chapters 7 and 8) were powerful pressures for reform in Bulgaria.

Subsequent reform throughout Eastern Europe has given greater responsibility to enterprise management, although not to the workforce. Rarely has reform allowed the market mechanism to set prices, to determine or to influence new investment. Only Hungary has pursued such market-oriented reform over a prolonged period. The more common strategy, applied most rigorously in East Germany, has been to rely on larger units of organisation.[1] These industrial associations have sought to promote efficiency through less central control over a smaller number of units. Higher management faces severe sanctions for poor performance. But even the continuing Hungarian reform, with its greater stress on profits and market prices as performance indicators and less ministerial administration, cannot fairly be called a full-scale repudiation of central planning. Only the economy of independent Yugoslavia, long outside the Soviet political orbit, has repudiated such planning in favour of what has been called market socialism. At least one Western economist now acknowledges the positive role of central planning as Eastern Europe's 'engine of growth' from the 1950s to the early 1970s, if not the past decade.[2] Few would dispute its role as 'an engine of structural change'.

Bulgarian confidence in the framework of Soviet-style planning has never been shaken, for whatever mixture of political motives and economic achievement. Official efforts to reshape that framework none the less began by the early 1960s. The enlarged Soviet-size collective farms created during the Third Five-Year Plan (1958–60) were the first target. Measures to streamline farm organisation and to give management more responsibility were soon extended, as we shall see, to industrial enterprises. For the next decade the Bulgarian strategy for reform essentially followed this approach, pioneered in East Germany. Little more than lip-service was paid to the need for profit indicators and market prices. By the late 1970s, however, a new series of measures made financial indicators more prominent and promised to apply market principles to some prices and new investment.

What sets Bulgaria apart form the Soviet Union is the virtually unbroken process of implementing, or at least discussing, economic reform for the past two decades. We must speak, according to Bulgarian economists, of reforms, rather than any one reform. One Western source that is generally critical of Bulgarian policy commends 'the degree to which its economic policy-makers have demonstrated that they are ready to undertake a new round of organizational changes' when the old approach is not working well; another acknowledges Bulgarian readiness 'to persevere with the reforms that have made it one of the most successful and economically progressive countries in Eastern Europe'.[3]

Bulgarian initiatives are always formally presented in terms of respect for Soviet precedent. Such initiatives have been too many and too widely implemented to be regarded as simply following Soviet leads. Economic reform in the USSR itself has been too sporadic and limited to support even the Western notion that Bulgaria has served as a laboratory for economic experiments which, if successful there, could then be introduced on Soviet soil. There has been too much Bulgarian flexibility and experimentation, at least compared to the reforms reluctantly introduced in the Soviet Union, to sustain this argument.[4]

The New System of Management, 1963–68

Post-war Bulgaria's first comprehensive set of reform measures was introduced in the period 1964–5 and endured without major alterations until 1968. This *nov sistem na rukovodstvo* (or New System of Management) had its origins in party discussions behind closed doors and also in public debate in scholarly journals during 1963. The green light for these deliberations may indeed have come from a *Pravda* article on 9 September, 1962, by the Soviet economist Evgeni Liberman. His ideas for a series of self-supporting, market-oriented enterprises to produce clothing and other consumer goods in the USSR had actually resulted in several successful pilot projects. His name became synonymous with Soviet economic reform for the next decade. Public discussion of possible economic reforms did first appear in East Germany and Czechoslovakia, as well as Bulgaria, following the Liberman article. Yet both the Bulgarian debate and the subsequent reform measures went well beyond the scope of Liberman's focus on consumer goods.

A scholarly debate encouraged by the Party's Central Committee agreed on the pressing need for a more comprehensive reform that would reach all branches of production. The decline in overall rates of economic growth during the early 1960s were contrasted with the still rapid rise in foreign trade. How could exports be maintained over time, a number of economists argued, if the growth of labour productivity in industry and construction, and eventually agriculture, slacked off? Petko Kunin, a rehabilitated supporter of the post-war economic chief Traicho Kostov, argued that state factories should now be self-supporting. Angel Mihailovski advocated the adoption of workers' councils on the Yugoslav pattern. Evgeni Mateev, perhaps Bulgaria's most eminent post-war economist, supported the majority's emphasis on new initiatives from management rather than labour. These discussions continued into 1966, when they reached their peak of intensity according to some Bulgarian scholars.[5] Yet no real agreement was ever reached on what part profits and market prices might play.

What one Bulgarian account recently called 'the great social experiment' began in June 1964.[6] Some 50 industrial enterprises, typically producing textiles or other consumer goods along the lines suggested by Liberman, were placed under the new system. Wages, bonuses and even funds for new investment were tied to plant profits, up to 70 per cent of which could be retained. Outside investment funds would come from bank credit, rather than the state budget. The number of compulsory targets for the Five-Year Plan was cut to four: physical output, upper limits on investment and on inputs, and foreign-trade targets. The pilot enterprises generally prospered, earning profits double the 5–10 per cent of gross income that was then the norm.

Despite these initial successes, internal party debate about how to introduce the system more widely without vitiating the planning apparatus delayed the publication of the reform's 'theses' until December 1965. By that time, however, enterprises responsible for 43 per cent of total industrial production were operating under this system. Their output increased 16 per cent during 1965, compared with 11.5 per cent for other enterprises. (Many of the former may well have been more successful firms in the first place.) By 1967, two-thirds of industrial production came from firms under the new system, which now went far beyond the bounds of consumer goods alone.

Its price regime was a compromise along the pattern evolving in Czechoslovakia. Retail prices were divided into three categories (fixed, flexible within limits, or free). Wholesale prices were supposed to be set along the same lines. In fact, most remained fixed. Producers' goods were still priced at average-industry cost plus 2 per cent profit, and consumer goods at cost plus turnover tax. Reform measures to allow 10 per cent mark-ups for high quality foundered on enterprise reluctance both to set prices for new goods and to wait until after retail sale to receive a mark-up on old goods.[7]

The Bulgarian retreat from this initial reform is typically linked in Western accounts to the chilling effect of the Soviet invasion of Czechoslovakia in August 1968. Building Soviet displeasure over the possible spread of Czech-style reforms probably played its part, but the timing and the rationale for the actual changes made in Sofia suggest an important Bulgarian side to the decision. Signs of recentralisation may be found as early as Todor Zhivkov's speech to the party plenum of April 1966. Discussions continued at subsequent party meetings. The decisions that were finally announced in July 1968 were probably made in some internal, preliminary form well before the end of 1967. These decisions eliminated the three-tiered price system. Free or flexible prices had never won wide acceptance in the party leadership. In the words of Grisha Filipov, when Planning Vice Chairman in 1964, 'The market can establish neither prices nor production . . . They should remain strictly according to plan.' In a speech in 1982, Filipov noted in retrospect that 'we had some difficulties because of the fact that there was no place from which to borrow a concrete mechanism corresponding entirely to our own concrete requirements and to our aims'.[8] The Soviet example, in other words, was of little concrete use for the New System of Management.

Helping to fuel resistance to reform were a series of corruption cases emanating from large enterprises. Certain directors had used flexible pricing and profit retention to line their own pockets. The scandals at a textile importing enterprise, which had landed the lucrative licence for bottling Coca-Cola, and a Black Sea shipping enterprise were only the most notorious.[9]

The several new planning indicators that were now added had two aims — to correct these abuses and to encourage exports for hard currency. Neither goal seems to fit the scenario of a Soviet crack-down on Eastern European reform, however viable this view may be elsewhere in the region. The new limits to the size of an enterprise's wage and salary bill, and to self-contracted obligations, were all aimed at curbing future corruption. Specifications for new investment did limit initiatives by enterprise management, but sought to maximize exports and to minimize imports.

The Bulgarian retreat of 1967–8, whatever its origins and extent, has tended to obscure two changes associated with the New System of Management. Both have become a permanent part of the economy's operation, one for agriculture and one for industry.

A new set of operating procedures was introduced on the enlarged collective farms in the early 1960s. This culminated in a complete set of farm statutes in 1967. In order to obtain better results from the large amount of new equipment introduced during the consolidation of the late 1950s, farms were assigned more agronomists and were paid higher prices for their produce. Their peasant members were no longer shuffled around between different tasks and sites, but placed in fixed brigades performing the same job as close to their village as possible. Directors and their agronomists still had to meet targets set by the Ministry of State Planning, but could now negotiate those targets at the farm, rather than send a representative to the ministry's forbidding building in Sofia. The increased labour productivity of farms adopting this set of changes undoubtedly helped to persuade the Bulgarian party leadership to extend the New System to industry in 1964–5. Most of its provisions had already been introduced on some collective farms by 1963. The farm statutes of 1967 ratified these provisions and also extended pensions, health benefits and other social services to peasant members before any other Eastern European regime. All of this prompted one often-critical Western observer to characterise Bulgarian agricultural policy of the 1960s by 'its willingness to imitate and experiment . . . [as] more pioneer than imitator'.[10]

For industry, the combination of some 2,000 industrial enterprises into 120 state economic organisations (DSO) in July 1963 seems at first glance only a step towards more concentrated central control. The effort to free these associations from ministerial supervision, beyond the New System's four indicators, admittedly foundered when more indicators were introduced in 1967. But it was the associations, not the ministries, that were charged with supervising the new system of supply contracts between enterprises. This system has grown steadily and now generates several thousand contracts a year, with prices fixed on the basis of enterprise bargaining, rather than ministerial fiat.[11] Allocating efficiency has clearly benefited, although by Western standards not as much as if a free market priced all supplies according to their relative scarcities.

First Financial Reforms

The Bulgarian turn toward financial reforms in general, not just rational pricing in particular, has been slow. Karl Marx regarded money as a facade behind which the real productive forces of labour power and capital goods were at work. Soviet economists have subsequently been reluctant to attach much importance to monetary policy. Post-war Bulgarian policy has shared this reluctance. Some initial monetary measures were taken during the 1960s, but more sweeping initiatives did not appear until the end of the 1970s. These are the subject of this chapter's final section (pp. 215–19).

The initial monetary measure of the 1960s was the exchange of all domestic currency on a 10–1 basis for a new 'heavy leva'. This 1961 exchange appeared at first as another instance of Bulgaria following a Soviet lead; in this case, it was the only Eastern European country to do so. Then came confusion about the gold purchase price of the new leva. The Bulgarian National Bank eventually set the price at 2.88 leva per gram, rather than the 1.32 rate first announced. This had the effect of devaluing the Bulgarian leva in terms of, if not in direct convertible exchange for, the 'hard' Western currencies by 40 per cent.[12]

Monetary controls have remained tightly held in the hands of the Bulgarian National Bank. Only the Bulgarian Foreign Trade Bank, established in 1964 with 40 million leva in capital, survived as a separate entity. The bank has probably endured because it grants domestic credits in hard currency as well as in leva, and because it has developed the ties to Western banks and their acceptances needed to overcome the inconvertibility of Bulgarian currency. In brief, it manages the country's balance of payments. The Bulgarian Investment Bank was merged with the National Bank in 1967, on the grounds that their separate procedures limited credit access and that a single centre for long and short-term credit would be more efficient. The Overseas Trade Bank opened in 1968 and the Industrial Bank and Agricultural and Trade Bank opened in 1969. None lasted more than two years. They lacked branches outside Sofia and were criticised for arbitrary judgement of credit applications.[13] The National Bank quietly absorbed their activities.

Bank credit in any case continued throughout the 1960s to constitute an insignificant share of investment funds. In 1964, such long-term credit amounted to just 6 per cent of enterprise investment. The state budget still furnished 58 per cent, and enterprise income the rest.[14] The main function of the National Bank was to furnish short-term credit.

Bulgarian fiscal policy reorganised the structure of budget revenues after 1960 with a view to trimming price distortion and to capturing more net income from large-scale enterprises whose unit costs of production were now declining. The brief experiment of the late 1950s, shifting expenditures in the unified state budget from the central toward the local level, was quickly abandoned.

Attention soon turned to the tax structure. Over half of total budget revenues had previously come from the turnover, or variable sales tax on consumer goods. That fraction was trimmed to one-third by 1968, and to one-quarter by 1982. These cuts reduced the aforementioned distortion in consumer prices set according to cost plus turnover tax. The place of this levy was taken by taxation of net enterprise income (and briefly, for 1964–6, by a tax on investment funds). The share of enterprise taxes in budget revenues rose from 13 per cent to 32 per cent between 1958 and 1970, and to 47 per cent by 1982.[15] Mitigating the higher rates of this new taxation were the lower tax brackets accorded to larger enterprises. This tax incentive prompted some enterprises to hire more workers, even beyond the reserve supply kept for 'storming' to meet targets at the end of the planning period. The incipient shortage of industrial labour was thus made more acute.

Evolution of Agro-Industrial Complexes

The more immediate labour shortage in agriculture prompted further combination of collective farms by the late 1960s. The party's ideological inclination towards the large scale also encouraged this step. The so-called agro-industrial complexes now came into being. Accompanying this major agricultural reorganisation was another Soviet-style propaganda campaign and a series of Bulgarian-style reforms aimed at specific problems arising from the reorganisation. Once again, as we shall see in the next section, these agricultural measures opened the way for a more comprehensive set of reforms, which also included industry.

The new agro-industrial complexes (APK) were from the start intended to accomplish more than simply the horizontal integration of existing collectives into state farms. This third stage of post-war Bulgarian agricultural development, following the creation and then the enlargement of Soviet-style farms, sought to industrialise agricultural production. The APK did not begin converting peasants into state farm employees earning a fixed wage until after 1972, on the pattern dominant in Soviet agriculture, but otherwise attempted to introduce industrial methods of production from the start. The effort began with modern facilities for mass egg-laying and livestock-fattening. Its final goal was vertical integration with food, fertiliser, or fibre-processing plants on the farm's own territory. This integration was to constitute a fourth stage in post-war agricultural development.

The party's Central Committee first discussed the new form of organisation, calling them 'agro-industrial combines', in its July plenum of 1968. Five experimental 'combines' were operating by the end of the year, and twelve more by the end of 1969. Several were merely mergers of existing state farms already organised on an industrial basis. They were hardly a major component in Bulgarian agriculture, with just 4 per cent of cultivated land. Others, however, combined several collective farms (TKZS) and introduced new methods of production. The 'combine' at Novi Kirchim, near Plovdiv, not only modernised livestock breeding and feeding, but also linked its fruit production to the canning-plant Vitamin. This vertical integration made it the first of what have since been called industrial-agricultural complexes (PAK).

Final party approval and comprehensive adoption of this new framework of complexes followed in 1970–1. The Central Committee set out the formal basis for establishing APKs or PAKs in its April 1970 plenum. By the end of 1971, all of Bulgaria's 744 collectives and 56 state farms had been merged into 161 complexes, mostly APKs. They averaged 24,000 hectares and 6,500 members. Average arable and fixed capital had both risen 4.6-fold, with grain and vegetable plots increased over 10-fold.

The reasons for the decision, according to the best Western analysis of the transition, go beyond the party's traditional Leninist faith in economies of scale and capital-intensive production.[16] The new complexes were also intended to boost the value-added component in Bulgarian exports by processing more agricultural goods. A further task was to raise the urban supply of food without diverting labour back from industry. Both of these goals were being frustrated by the relatively poor agricultural performance of the late 1960s. As noted in Chapter 7, meat, fruit and vegetable production had lagged in particular. According to one Bulgarian account, canning sustained 'unbearable losses'.[17] Popular discontent at seeing these goods in abundance only before major holidays or at the Plovdiv International Trade Fair could not be denied, especially among those aware of the foodstuffs regularly available in neighbouring Yugoslavia and Greece by the 1960s. Party apprehension rose. Then, in 1972, Todor Zhivkov made a major speech in which he promised a ten-year programme to raise living standards in general, and to bring food consumption in particular, up to the 'scientific norms' set by the United Nations (see Table 8.5).

The new agricultural framework afforded only mixed results during its first years of operation. Yields per hectare rose between 1972 and 1976 for wheat, apples and grapes, but not for the other crops which, together, made up a majority of cultivated output. Gross agricultural output, including livestock, did increase by an annual average of 5.3 per cent for 1972–6, compared with 3.1 per cent for 1966–70. At the same time, gross investment in fixed capital still grew at a slightly higher rate, 5.5 per cent, than did output. Investment growth was at least lower than the 6.9 per cent average for 1966–70. The anticipated freeing of more farm labour for industrial jobs, however, did not take place. The annual reduction in agricultural employment dropped from 4 per cent to 2 per cent. Productivity of the remaining farm labour also declined. Agriculture's share of gross investment in fixed capital fell back to 18 per cent by 1976, a level last seen in the mid-1950s. Investment in human capital also slipped. The number of specialists entering agriculture dropped from 6 per cent to 2 per cent of the national total.[18]

In the absence of funds to stem the declining level of investment, more reorganisation has been the only Bulgarian recourse. One change has been to follow the logic of reducing the number of units for control to its final extreme. In 1976, a single national Agro-Industrial Union was set up. The Ministry of Agriculture and Food, itself a recent merger of two ministries, was initially in charge. Although elections from the various complexes were held and connections from scientific institutes to the national union were created, neither device could overcome the new ministry's heavy-handed surpervision of co-ordinated activities. Press criticism in 1979 of the ministry's 'petty tutoring and rude interference' signalled the party's growing displeasure. The ministry was thereupon abolished, and the union charged with all of its functions.[19] The change has at least eliminated the wasteful assignment of many Bulgarian scientists to part-time duties at one complex or another.

More positive results have come from the slimming down of the complexes themselves. By 1977, the APKs totalled just 143. Several were larger than 100,000 hectares and 25 were over 36,000 hectares. Their reduction in size and the creation of new, smaller complexes began the following year. By 1982, the total of old and new APKs reached 296. As a result of this doubling of numbers, the average size has been cut in half, to 16,000 hectares. The management of each complex, typically a merger of five or six farms, is now divided into directorates for each type of production. The intermediate level of management between the directorate and the working brigade was swept away. Most importantly, the number of annual indicators for plan fulfilment, which had ballooned to fourteen, fell to four. Surviving targets are obligatory deliveries, payments due to the state budget, a minimum for export earnings in hard currency, an import maximum, and a maximum of inputs from other sectors.[20] Greater APK freedom to negotiate prices on production beyond the plan and to sign contracts for its own supplies is therefore provided.

The other major change since the creation of the complexes has been the expansion of two other types of organisation, the industrial-agricultural and scientific-productive. These PAKs and NPKs (or NPOs) are intended to achieve vertical integration and introduce advanced technology by bringing processing plants and scientific institutes directly into the complex. In fact, industrial enterprises have tended to dominate the management of PAKs, and the institutes the NPKs. Both sorts of complexes have also been substantially larger than the APKs, over twice the arable land and almost twice the labour and fixed capital.[21] Most of them have been assembled by combining several APKs with existing enterprises or institutes.

The first PAK was the Bulgarian Sugar Association. Set up in 1972, Bulgarski Zakhar encompassed eight APKs and seven refineries by 1976. Its vast cultivation of sugar beet demanded fodder crops for rotation. Livestock herds were added to make use of the fodder. Sugar-beet yields did not generally increase, however. With the major exception of the Rodopa meat-exporting enterprise, which had already integrated with several collective farms in the 1960s, results from the other PAKs were disappointing. Capital/output and labour/output ratios were soon at the same high level of the APKs. Hence the turn to the institute-dominated NPKs and NPOs.

The first NPK, or scientific-productive complex, was organised in 1974 for viticulture and wine production. Wine has subsequently proved to be one of the most successful new hard-currency exports from Bulgaria over the past decade (see Chapter 8). That same year, the first of the more frequent and more loosely structured NPO 'organisations' (associations in Western terms) was put together for poultry production, another area of subsequent success. The largest of them has been the merger of two Plovidiv institutes, the two huge and already profitable APKs at Purvenets and Novi Kirchim, with additional vegetable and fruit areas, and with the local hothouse facilities of the Bulgarplod export enterprise for fruit and vegetables. Its employees and members total 32,000. Set up in 1977, after a year's delay for negotiations, this NPO has increased labour productivity and cut production costs significantly, though not as much as planned.[22]

The Rise of Personal Plots

Although some of these further combinations among the agricultural complexes of the socialist sector have boosted agricultural exports, family farming for private profit has been largely responsible for the improved supply of domestic foodstuffs during the past decade. This production, it must be emphasised, has not come from privately *owned* land. The completion of collectivisation cut the share of private land to 2 per cent of the arable total by 1959. The figure has only declined further since then, touching 0.4 per cent by 1972. Most of this tiny fraction was located in mountainous areas.

Instead, the recent revival of private farming in Bulgaria has come about on plots leased from the agricultural complexes. These are the so-called *personal* plots that were first authorised to peasant households in 1957, as compensation for joining the collective. Based on the system of plots granted to Soviet *kolkhoz* households since 1935, their size varied by land use 1–10.2 hectares for intensive cultivation, 0.5 hectares for grain, and 1 hectare for mountainous areas or fields not used for large-scale cultivation.[23] Another 0.1 hectare was added for each head of cattle raised. Regulations in 1963 and 1967 reaffirmed the permanence of these plots, which typically adjoin the family house. Together, they amounted to about 10 per cent of cultivated land.

Special measures to expand the number, activities and market sales of the personal plots did not begin until 1971. That year saw previous restrictions on the number of livestock lifted. Major encouragement for a variety of marketed production materialised in 1974 and 1977. The party leadership had, like that of Hungary, overcome the ideological reservations that continue to restrict the Soviet plots. From 1974 onwards, peasant households have been permitted to lease additional plots and to have free access to fertiliser, fodder seed and equipment, from the agricultural complex. The Central Co-operative Union, cut off from the complexes after 1970, was now empowered to sign contracts for the sale of plot surpluses at their network of town market-places. The Union has also provided a growing number of delivery points and received rights to inspect leased plots and to set delivery targets. To encourage new takers, loans were extended and income taxes waived for 1974. More importantly, delivery prices were raised. This was also the year in which the reduction of Bulgaria's non-agricultural working

Table 9.1: Cultivated Area of Personal Plots, 1965–80 (in 1,000 hectares)

	Grains		Vegetables and fruit		Forage crops	
	Area of plot	Percentage of total	Area of plot	Percentage of total	Area of plot	Percentage of total
1965	302	12.7	46	31.2	41	5.4
1970	257	11.3	42	27.5	44	6.5
1975	299	12.6	46	29.2	48	7.0
1980	287	13.0	65	41.3	56	6.4

Sources: *Statisticheski godishnik na NR Bulgariia, 1972* (Sofia, 1973), p. 201; *1982*, pp. 268–9.

week, from 48 to 42.5 hours, was largely completed. Townspeople could now enjoy a Western-style weekend. They were free to assist their typically older relatives in the countryside with personal plots.[24]

Further measures in 1977 actively promoted urban participation. Town residents might themselves apply for plots directly to the complex, even if they had no relatives there. (More recently, complexes have taken to advertising available plots in newspapers, complete with number to telephone for further information.) Applicants need only be already employed in the socialist sector and refrain from using their plot to build a summer-vacation house. In addition, no lessee can employ hired labour, in effect non-family members, or leave his or her job in the socialist sector, whether in a factory or on the complex, to work full-time on the plot.[25] Size limits were also removed at this time, and further tax exemptions and fodder allocations offered. A flood of applications for the additional 150,000 hectares of land offered from the agricultural complexes brought the amount of land under personal plots to 14 per cent of the cultivated total by 1980.

Accompanying the increase in the area of personal plots between 1965 and 1980 was a change in crop structure, which favoured urban market-places. The rise in the plot shares of vegetable and forage crops (the latter for raising livestock) more than made up for the decline in the area devoted to grain noted in Table 9.1. By 1978, personal plots accounted for 22 per cent of Bulgarian vegetable and feed-grain production. The proportions were even higher for fruit, 39 per cent, and more for individual vegetables such as green peppers and green beans. Yields were growing more quickly for wheat and forage crops than in the socialist sector, surpassing those levels by 1981. That year overall crop output from the plots grew by 4 per cent, compared with 3.5 per cent from the socialist complexes. By 1982, the plots produced 33 per cent of all vegetables and 51 per cent of potatoes.

Animal products have done even better. Meat and milk production from the plots jumped almost threefold from 1973 to 1975. Since then they have consistently provided nearly 40 per cent of meat sales, 30 per cent for milk, and over one-half of egg production. All together, by 1982, the plots accounted for one-quarter of agricultural output and of farm workers' income.[26]

For the urban consumer, deliveries from the plots have clearly become indispensible for all major foodstuffs except bread. Food prices were increased again in 1979, and are high by Eastern European standards. The more successful extended-family enterprises earn profits that are large enough to pay for luxuries like second cars or lavish weddings. Townspeople sometimes take resentful note of these trappings. At the same time, plot sales to town markets resolve the paradox of overall rates of agricultural growth, which have been low during the past decade, and an urban food supply which has undeniably improved.

Industrial Concentration during the 1970s

The latest round of combining existing enterprises into a smaller, and presumably more easily controlled, number of units spread quickly from agriculture to industry. What a West German economist has called 'an extremely consistent process of concentration' was again at work in industrial organisation by the mid-1970s.[27] Recent official views seek, in retrospect, to link such concentration with a parallel emphasis on greater responsibility for individual managers and brigades. At the time, however, this emphasis was greatly overshadowed by the initial promise of the Soviet campaign to use automated systems and information processing to boost productivity. Soviet planners had launched the campaign in 1968. It was an alternative to the Czech proposals for decentralising economic reforms that had emanated from the soon-suppressed Prague Spring. The Bulgarian Council of Ministers established a potentially powerful committee to replace the industrial ministries in co-ordinating production with long-term plans. Little specifically came of its activities, however. The main legacy of the cybernetic approach has been a more comprehensive system of collecting economic data.[28]

The basic Bulgarian strategy for better management in the early 1970s none the less was wedded to promises of automated management under a simplified surperstructure. Measures followed to reduce the number and increase the size of the associations into which industrial enterprises had already been combined in 1963. The enterprises were themselves larger than their Western counterparts. They dwarfed the tiny firms, which were identified in earlier chapters as a constraint on modern industrial development before and immediately after the Second World War. By 1970, plants with over 1,000 employees (one-quarter of which were located in the wider Sofia area) accounted for 47 per cent of the labour in the country's 2,471 industrial enterprises. Here was a higher level of concentration than that indicated by East German or Western European plant size.[29] The number of enterprises was less than before the Second World War (see Table 3.5) and under one-half the number in excess of 6,000 on the eve of nationalisation in 1947 (see Chapter 5).

At the end of the 1970s, the number of state economic organisations (DSOs) into which industrial enterprises had been grouped since 1960 was virtually cut in half. The 64 enlarged organisations received the right to make decisions for their enterprises about new investments, bank credits or budget subsidies. The larger enterprises (now called subsidiaries) could still sign their own supply contracts and maintain their own bank accounts, but ceased to be legal entities. Smaller enterprises (now called subdivisions) became fully dependent on the enlarged DSOs. By 1975, moreover, seven new industrial ministries had set up eleven 'combines', one for each major branch of industry. Praised by Todor Zhivkov as the organisational 'unit of the future', which would finally achieve vertical integration, the combines effectively bypassed by DSOs. The latter retained only a formal identity, and the chemical industry dissolved even that.[30]

The advantages of this streamlined organisation appeared to be several. It again promised greater economies of scale — the grand passion of Soviet (and much American) investment strategy — through easier specialisation and a simplified flow of information. The combines would presumably be better equipped to make investment decisions than either the small number of old ministries or the large number of enterprises. The same superior co-ordination was expected in distributing supplies of scarce inputs, labour included. The new framework was to introduce the same sort of linkage between specific industrial enterprises and scientific institutes that had been arranged for the agricultural complexes. Finally, the combines were relied upon to improve the effectiveness of 'counterplanning', the major Soviet innovation of the early 1970s. Bulgaria adopted this device in 1971. Counterplanning called on enterprises to submit their own, presumably higher targets for obligatory production, as a response to those set by the State Planning Commission. Such counterplans would presumably draw down input reserves, whose hoarding held production under three-quarters of capacity in many industrial enterprises.

The results of these decisions to reduce and then to eliminate the DSOs in favour of eleven combines appear to have been disappointing.[31] Counterplanning continued to leave reserve industrial capacity unused and did nothing to improve quality control. Both Western buyers and domestic customers remained unsatisfied with the quality of too many Bulgarian manufactures. New planning indicators that set norms for cost reduction actually reduced quality in a number of cases. The programme mandating that every scientific institute assign its members to part-time service for a specific enterprise did not work any better for industry than it had for agriculture. Individual members of institutes lacked the power to reach combine or ministry levels, where decisions to import or to invest in new technology were made.

In summary, the new framework only accentuated the dangers of socialist monopoly. These dangers were divorced from the abuses of exorbitant pricing in earlier capitalist monopolies, but not from the risk of restricted output or irrational decisions immune from punishment by competitors in the market-place. Bulgarian authorities, from Todor Zhivkov down, have refused to acknowledge explicitly the threat of monopoly practices, which are associated exclusively in their vocabulary with capitalist economies. But the monopolistic abuses of irrational decisions taken arbitrarily at the top, or rational ones poorly implemented at the enterprise level, were increasingly criticised in party meetings and the press. By the end of the 1970s, a further set of reforms had therefore been launched.

Financial Reforms since 1978

The latest Bulgarian reforms began in 1978 and have continued under the designation, since 1979, of the New Economic Mechanism (NEM). This time they have involved the agricultural complexes and the industrial enterprises almost simultaneously. Their principal measures, in 1978–9 and 1982–3, have been financial incentives and accounting regulations aimed at all levels of management, but especially at the smallest unit of labour, the brigade. The overriding aim has been to provide financial accountability. The absence of such accountability, according to a leading Bulgarian economist, has kept the concentrated organisation of and linkages between Bulgarian enterprises from having the co-ordinating effect that they do in a large Western corporation.[32] With this co-ordination, the Bulgarian economy can aspire to become a single, socialist corporation. The means to this end are seen as those of the financial market-place. In the words of Todor Zhivkov, the Bulgarian economy should operate in future on the basis of 'the money-goods relationship' and its indicators: prices, profits, interest rates, credit, etc.

Two important initiatives were launched in 1978. A series of decrees lifted the long-standing limits on enterprise investment, previously set by the State Planning Commission. In their place was put an investment plan based on the enterprises' contractual obligations and credits with the Bulgarian National Bank. The bank would also monitor the cash balance of enterprise contracts with customers and suppliers, granting credits only when required. Self-finance thus became the watchword for all economic organisations, according to the best Western summary of the 1978–9 reforms.[33]

The second major change made wages and salaries a residual, rather than a first claim on gross enterprise income. Wages could rise only after an increase in labour productivity, and then by half of that increase. Management salaries could moreover be cut by as much as 20 per cent, if the complex or enterprise failed to meet its norms for production and productivity. Since their introduction, the formula for these sanctions against management salaries has changed several times. The consistent aim has been to make them fearsome. In 1980–1, a fraction was actually withheld pending the enterprise's successful performance; more recently, that fraction must be repaid by the manager following an unsuccessful performance. Here, according to one American appraisal of the latest

Bulgarian reforms, is the biggest change made so far.[34]

The New Economic Mechanism was formally proclaimed in 1979, first for agriculture and foreign trade, and shortly thereafter for industry. Its principal concern was to redefine the norms for successful performance and also for the disposition of profits. Binding criteria for performance were limited to five financial indicators for agricultural complexes and industrial associations, and to four for individual enterprises.[35] Profit criteria were set only for the complexes or associations. Three separate funds for reinvestment received first claim on net income. One, the so-called Additional Material Incentive Fund, was specifically intended to compensate firms for avoiding excess cost reduction and the poor quality that was often a by-product. A new set of proportional taxes would then take an average of 80 per cent of net enterprise income for the state budget, thus preserving the latter's economic leverage. Although budgetary subsidies were not being eliminated, the NEM directives repeatedly stressed the responsibility of higher management and even individual brigades for losses incurred. Complexes or associations were given explicit freedom to sign their own contracts with suppliers and customers at home and abroad.[36]

Several supporting measures were also introduced during the late 1970s. A campaign to cut the excessive number of administrative and supervisory personnel, one of a series in the post-1950 period, did not achieve its ambitious target, but never the less trimmed management's share in the labour force from 13 per cent to 9 per cent. In 1979, as noted in Chapter 8, prices of many consumer goods were raised by 25–30 per cent, and wages or salaries just 10–15 per cent, in order to offer producers a greater chance for profitable production.

The Bulgarian National Bank (BNB) was already allocating more investment credit, a rise from 7 per cent to 39 per cent of the total distributed between 1965 and 1970, according to Table 9.2, at the expense of the state budget's share. But the BNB's decisions remained too dependent on the permissive preferences of planning authorities in Sofia. In order to combat this frankly admitted 'centralism', the bank was granted some flexibility in restricting its terms and in charging interest rates above the nominal 2 per cent still assessed. All these measures were designed to promote greater rewards for efficiency and to reduce the number of unfinished, often unprofitable new projects. The latter had accounted for 57 per cent of all Bulgarian investment as late as 1976. By 1981, this

Table 9.2: Sources of Economic Investment, 1956–81 (percentages)

	1956	1960	1965	1970	1975	1981
Enterprise funds	29	23	30	34	25	45
Bank credits	9	13	7	39	54	28
Budget subsidies	62	64	63	27	16	28

Source: *Statisticheski godishnik na NR Bulgariia, 1967*, p. 321; *1982*, p. 466.

'new' proportion had been cut in half, to just 27 per cent.[37]

Perhaps the clearest institutional commitment to increased efficiency has come from the creation, in April 1980, of the Bulgarian Industrial Economic Association (BIEA). Established outside the ministerial framework, but with strong backing from the party's Central Committee, the BIEA is a voluntary organisation of existing industrial associations and enterprises. Its president holds a higher rank than any economic minister does. Its membership represents over 700 organisations, accounting for 75 per cent of industrial output. Members retain their previous autonomy, although the BIEA is a legal entity. Its sole purpose has been to increase member efficiency in domestic and foreign markets. To this end, it has set up its own computer centre to analyse member income, has helped to sell surplus inventories, and even helped to co-ordinate regional construction projects.

The BIEA's most successful project to date has been the promotion of some 200 new, smaller enterprises, designed to produce either consumer goods or component parts that are in heavy demand. They remain state enterprises, unlike their famous Hungarian counterparts, but their size (less than 200 employees, in fact most have under 50) and product mix are quite similar. The list of best-selling products runs from denim jeans, jogging shoes and babywear, or scarce components such as zippers on the domestic market, to ceramics and oriental rugs for export. Plans were announced in 1984 to boost the number of such firms to 500 and to accord them 10 per cent of funds for new investment.[38] Yet the very success of these new, smaller enterprises has highlighted the failure of the larger complexes and associations to perform as well under the NEM. The youthfulness of a number of the small-enterprise directors also calls attention to the need for more rapid advancement of able young managers elsewhere in the economy.

Already in 1982, both Bulgarian economists and the party leadership were prepared to admit that the NEM had not led to the anticipated upturn in overall productivity and efficiency. Aggregate economic growth, as detailed in Chapter 7, had slowed to its lowest post-war level by the early 1980s. The limits on better Bulgarian performance, moreover, seemed to go beyond the international barriers of costlier oil supplies and slower Western and Soviet growth. The Bulgarian response has been to begin a second set of reforms under the rubric of the NEM, with the new promise that more will follow, if these do not yield satisfactory results.

The 1982–3 measures have concentrated almost exclusively on financial incentives and that most change-resistant of Bulgarian indicators, the price mechanism. 'Final economic performance', that is, net income, has been identified as the major basis for judging plan fulfilment. The only other targets are tax payments, maximum-use levels for domestic and imported inputs, and minimum export levels. The emphasis on self-supporting net income has been extended downwards to the brigade-level and upwards to the large associations headed by economic branch ministries. The guarantees of a minimum wage have been removed for workers, as well as management at all levels. A pilot project in Ruse is experimenting with ranking managers' performance over five years. A substandard record by their enterprise, that is, net income ranking in the lowest 10 per cent of the pilot group, will supposedly result in demotion. Ministers themselves are now subject to salary reductions if their industrial association fails to meet the streamlined list of targets. They are also held responsible if their massive associations meet targets by using monopoly power to hoard inputs etc. Their access to budgetary subsidies for new investment is now rare, and then for a fixed term only.[39]

Most investment capital outside net income must be sought from the Bulgarian National Bank. Its increasingly independent guidelines include the authorisation to hold regional competitions for investment funds. Interest rates remain low, at 2.5–8 per cent. But the competitions may be described as auctions of the sort that Yugoslavia attempted in the mid-1950s. Only a minority of applicants can win in any one competition.[40] The regional distribution may also help to spread the bank access from Sofia to the provincial towns. Overcentralisation in Sofia is a problem that has plagued the National Bank's operations, we may recall, since before the First World War.

The State Agency for Prices and Price Information has been

given a new mandate and new data-processing equipment to collect cost data independently. It continues to police the fixed prices set for consumer necessities and some industrial inputs by the Council of Ministers. But for the majority of prices, the Agency is charged simply with setting maximum levels under which the various economic associations, complexes and subunits may negotiate prices with each other.[41] Its guidelines for these prices stipulate only that they must prevent either excess profits or essential production at a loss.

The effectiveness of this latest and most serious attempt to restructure prices, so that they better reflect production costs and world prices, deserves to be watched closely. Its fate may determine whether the prolonged Bulgarian experience with economic reform will finally stimulate the economy's aggregate performance, rather than only that of certain individual enterprises. That fate may depend, in turn, on whether thorough price and economic reform will finally spread to the Soviet Union, which by the weight of its trade alone sets important parameters for the Bulgarian economy. The brief tenure of Yuri Andropov as Soviet leader in 1983–4 seemed, from the Bulgarian viewpoint, to promise such reforms. In the event, only the same temporary tightening of labour discipline that Andropov launched in the USSR spread to Bulgaria before his death.

Notes

1. An instructive set of articles on all the Eastern European economies, including the Soviet Union, may be found in Alec Nove, Hans Hermann Höhemann, Gertrud Seidenstecher (eds.), *The East European Economies in the 1970s* (London: Butterworths, 1982).
2. See Nita Watts, 'Eastern and Western Europe', in Andrea Boltho (ed.), *The European Economy: Growth and Crisis* (Oxford: Oxford University Press, 1982), pp. 284–5.
3. Radio Free Europe Research (RFER), *Bulgarian Situation Report*, 12, 2 July 1982; *The Economist* (London), 14–20 January 1984, p. 13.
4. On Soviet reform, see Alec Nove, 'The USSR: Economic Policy and Methods after 1970', in Nove *et al.*, *East European Economies in the 1970s*, pp. 17–43.
5. These articles appeared primarily in the journals *Novo vreme, Ikonomicheska misul* and *Planovo stopanstvo*. Some are translated in the journal *East European Economies* for these years. The lines of argument are summarised in J.F. Brown, *Bulgaria under Communist Rule* (New York: Praeger, 1970), pp. 161–7.
6. Ivan Nikov, *Ikonomicheskiiat podkhod* [The Economic Approach] (Sofia, 1983), pp. 50–4. In Bulgarian terms, the 'economic approach' stands for the objective search for economic efficiency, in contrast to the earlier reliance on subjective administration for political control.

7. George W. Feiwel, *Growth and Reforms in Centrally Planned Economies: The Lessons of the Bulgarian Experience* (New York: Praeger, 1977), pp. 72–5, 97–127.
8. Brown, *Bulgaria*, pp. 167–72; *Rabotnichesko delo*, 15 January 1982. Also see Michael C. Kaser, 'The Industrial Enterprise in Bulgaria', in Ian Jeffries (ed.), *The Industrial Enterprise in Eastern Europe* (New York: Praeger, 1981), pp. 85–6.
9. Heinrich Vogel, 'Bulgaria', in Hans-Hermann Höhemann, Michael C. Kaser and Karl C. Thalheim (eds.), *The New Economic Systems of Eastern Europe* (Berkeley, California: University of California Press, 1975), p. 215.
10. Brown, *Bulgaria*, pp. 209–13.
11. RFER, *Bulgarian Situation Report*, 20, 26 January 1982. This contractual system is similar to Hungarian practice, and in sharp contrast to the continuing Soviet reliance on the planning mechanism and the Grossnab central agency to distribute inputs to enterprises. See Nove, 'The USSR', pp. 37–43.
12. The dollar rate remained at 1.17 leva = $1 until appreciating to 0.85, in tandem with the West German mark in the late 1970s, compared with 6.80 leva for 1951–61, or 0.68 after a 10–1 conversion. Brown, *Bulgaria*, pp. 144–5; Nikolai Todorov et al., *Stopanska istoriia na Bulgariia, 681–1981* [The Economic History of Bulgaria, 681–1981] (Sofia, 1981), pp. 518–22. By 1983 the dollar rate had fallen to 1.34, again in tandem with the mark. Tourists have received a 50 per cent premium since 1974.
13. T. Totev, 'Bankovata sistema na NR Bulgariia', *Finansi i kredit*, vol. XXXI, no. 7 (1981), pp. 48–50. On the Yugoslav banking system, see Laura D'Andrea Tyson, *Yugoslav Socialism: Economic Performance in the 1970s* (Berkeley, California: University of California, 1980).
14. George W. Feiwel, 'Economic Reform in Bulgaria', *Osteuropa-Wirtschaft*, vol. 24, no. 2 (1979), p. 80.
15. The local share of budget expense fell quickly back from its 1959 high point of 30 per cent to the previous level of 17 per cent. In the absence of detailed breakdowns of budget expenses and revenues in official statistical publications since the late 1960s, we may note only that aggregate expenses rose from 52 per cent of national income (NMP) in 1960 to 58 per cent by 1970, and to 63 per cent by 1980. Over one-half of these funds have until recently been directed toward investment in the economy, and about one-third toward social services. By 1980, however, the former had fallen to 44 per cent and the latter risen to 42 per cent. The military share increased slightly in the early 1970s, from 5 per cent to 6 per cent of budgeted expenses. Although it was exceeded in Eastern Europe only by East Germany, three-quarters of this relatively large defence share is spent on military salaries. Expenses for military research and development, never included in Eastern European defence budgets, are presumably small for Bulgaria. The economy benefits to the considerable extent that many units are regularly assigned to non-military construction projects, on the inter-war pattern established by Stamboliiski's *trudovak* battalions (see Chapter 2). Feiwel, *Growth and Reforms*, pp. 99, 115, 145; N. Popov, *100 godini na bulgarska ikonomika* [100 years of the Bulgarian Economy] (Sofia, 1978), p. 313; annual budget reviews by the Finance Minister, V. Velchev, in *Finansi i kredit*, vols. 21–33, no. 1 (1970–82), pp. 3–12. Vienna Institute for Comparative Economic Studies, *Comecon Data, 1979* (New York: Holmes and Meier, 1979), pp. 35, 399. On military expenditure, see Thad P. Alton, Gregor Lazarchik, Elizabeth M. Bass and Wassyl Znayenko, 'East European Defense Expenditures, 1965–1978', in Joint Economic Committee of the US Congress, *East European Economic Assessment*, pt 2 (Washington, DC: US Government Printing Office, 1981), p.. 409–33.
16. Paul Wiedemann, 'The Origins and Development of Agro-Industrial Complexes in Bulgaria', in Ronald A. Francisco, Betty A. Boyd and Roy D. Laird (eds.), *Agricultural Policies in the USSR and Eastern Europe* (Boulder, Colorado:

Westview Press, 1980), pp. 98–109.

17. P. Daskalov in *Ikonomicheska misul*, vol.1 (1969), pp. 22–34, as cited in Wiedemann, 'Origins', p. 107.

18. Ibid., pp. 122–31.

19. Todorov et al., *Stopanska istoriia na Bulgariia*, pp. 565–70. Also see RFER, *Bulgarian Situation Report*, 34, 14 February 1977.

20. Everett M. Jacobs, 'The Impact of Agro-Industrial Programs in East-European Agriculture', in Francisco, Boyd and Laird (eds.), *Agricultural Policies*, pp. 238–41; Organization for Economic Co-operation and Development (OECD), *Prospects for Agricultural Production and Trade in Eastern Europe*, vol. 2, *Bulgaria* (Paris: OECD, 1982), pp. 142–4. On the redistricting of Bulgarian territory into 'inhabited places', see RFER, *Bulgarian Situation Report*, 3, 15 February 1979.

21. Wiedemann, 'Origins', pp. 111–20.

22. Jacobs, 'Impact of Agro-Industrial Programs', pp. 252–3; RFER, *Bulgarian Situation Report*, 2, 21 January 1977 and 34, 14 February 1977; Todorov et al., *Stopanska istoriia na Bulgariia*, pp. 564–70.

23. Brown, *Bulgaria*, pp. 207–8; RFER *Bulgarian Situation Report*, 12, 17 January 1975. Soviet agricultural reforms have centred on the spread of state farms and wage-labour thereon. On the evolution of Soviet agriculture, see Karl-Eugen Wädekin, *Agrarian Policies in Communist Eastern Europe* (The Hague: Martinus Nijhoff, 1982), pp. 14–30, 44–62.

24. OECD, *Prospects*, vol.2, pp. 177–80; RFER *Bulgarian Situation Report*, 30, 23 November 1977.

25. Abuses have none the less appeared in the use of personal plots, for example, resale of the collective's livestock, excessive use of collective equipment and supplies, excess profits and even the abandoning of jobs in the socialist sector. See Mark Allen, 'The Bulgarian Economy in the 1970s', in Joint Economic Committee of US Congress, *East European Economies Post-Helsinki* (Washington, DC: US Government Printing Office, 1977), pp. 683–5.

26. *Statisticheski godishnik na NR Bulgariia* [Statistical Yearbook of the PR Bulgaria], *1975* (Sofia, 1976), pp. 219–36, *1983*, pp. 262–76; OECD, *Prospects*, vol. 2, pp. 177–80; RFER, *Background Report*, 224, 28 December 1984.

27. Vogel, 'Bulgaria', p. 217.

28. Ibid., pp. 207–11.

29. The 98 largest enterprises employed 23 per cent of Bulgarian industrial labour and were responsible for 75 per cent of industrial capital and 77 per cent of profits. Total enterprise numbers for industry dropped by 13 per cent for 1970–7, thanks to a two-thirds decline by artisan co-operatives. Their share of industrial production dropped from 10 per cent to 5 per cent. Todorov et al., *Stopanska istoriia na Bulgariia*, pp. 461, 489–97, 540–58.

30. George W. Feiwel, 'Economic Development and Planning in Bulgaria in the 1970s', in Nove et al., *East European Economies in the 1970s*, pp. 226–37.

31. Ibid., 238–48; Feiwel, 'Economic Reforms', pp. 75–86. Alec Nove aptly calls this Soviet-style approach 'the curse of scale', in his 'USSR', pp. 30–1.

32. Author's interview with Academician Evgeni Mateev, Higher Economics Institute Karl Marx, 10 May 1984.

33. Kaser, 'Industrial Enterprise', pp. 88–9.

34. Marvin R. Jackson, 'Recent Economic Performance and Policy in Bulgaria', Joint Economic Committee of the US Congress, *East European Economies: Slow Growth in the 1980s*, pt 3 (Washington, DC: US Government Printing Office, 1985), p. 25. The present author shares this view.

35. Kaser, 'Industrial Enterprise', p. 88; Todorov et al., *Stopanska istoriia na Bulgariia*, pp. 582–5. Compare these few indicators with the large number of

remaining Soviet targets noted in Nove, 'The USSR', pp. 35-43.
  36. Feiwel, 'Economic Development', pp. 226-37.
  37. Ibid., pp. 238-40; Feiwel, 'Economic Reforms', pp. 76-80; Todorov *et al.*, *Stopanska istoriia na Bulgariia*, pp. 577-81.
  38. RFER, *Bulgarian Situation Report*, 12, 23 July 1982; 1, 16 January 1985.
  39. Jackson, 'Recent Economic Performance', pp. 28-31; RFER, *Bulgarian Situation Report*, 20, 26 January 1982, summarising a lengthy speech on the NEM by the Prime Minister, Grisha Filipov, printed in full in *Rabotnichesko delo*, 15 January 1982.
  40. Author's interview with Ivan Dragnevski, Vice Director, Bulgarian National Bank, 16 May 1984.
  41. Nikov, *Ikonomicheskiiat podkhod*, pp. 148-56; Cam Hudson, 'Bulgarian Economic Reforms: Between the Devil and the Deep Black Sea', RFER, *Background Report*, 142 (Bulgaria), 2 July 1982.

CONCLUSION

The prospects and problems facing the Bulgarian economy during the rest of the twentieth century derive principally from the socialist framework put in place by the post-war Communist government. More distant historical dimensions must of course be added, if we are to grasp the full relevance of Bulgaria's economic past to its future. But we cannot honestly conclude that they deserve pride of place over the major post-war changes described in the past five chapters. Soviet-style central planning of production and investment, the Marxist priority of rapid and balanced industrial growth, the ties to the other socialist economies, primarily the USSR, and the Bulgarian Communist Party's persistent efforts to reform its own system since 1960 — these are the major facts of the country's present economic life.

Investment strategy continues to emphasise heavy industry. To the extent that this emphasis promotes an increasing array of electrical equipment exportable even to Western markets, it is plainly positive. So was the construction of electric power plants during the 1950s, which first made feasible the nationwide spread of modern industry. On the other hand, the enduring Bulgarian determination to manufacture most of its own ferrous metallurgy, despite a record of inefficient production and the need to import iron ore and coking coal, does not hold out much promise. The delayed conversion to nuclear energy, now aiming to provide half of all electric capacity by the year 2000, will at least relieve some of the drain on remaining coal reserves.

Investment strategy continues to neglect light industry. The principal victims of this neglect throughout the post-war period have been consumer goods, mainly for domestic use, and processed foods for export. The spate of new small firms charged with producing better-quality consumer goods have now recorded some early successes. More investment and more Western imports will probably be required, however, if such goods are to be exported profitably outside of Eastern Europe. Food-processing plants are in urgent need of modernisation. If their production is to meet world standards, new Western processing equipment must be imported. Otherwise, the proportion of processed food in

Bulgarian exports will continue to fall.

If the Eastern European Council for Mutual Economic Assistance were pursuing full-scale integration of its member economies, the share of Bulgarian food exports would undoubtedly have to rise again. But CMEA's principal goal appears instead to be the convergence of its members' productivity levels in all industrial branches. Hence its support for Bulgaria's new metallurgical complex at Burgas. In addition, the USSR and Eastern Europe now promise to be a tougher market for Bulgarian manufactured exports and a more expensive source of raw materials, especially of Soviet oil. If, in return, the Eastern European economies could guarantee Bulgarian enterprises a supply of industrial equipment at the highest levels of current technology, the genuine Bulgarian advantage enjoyed in past economic relations with CMEA might be more certain to continue.

In addition to CMEA ties and the new socialist framework, the course of the post-war Bulgarian economy stands separate from the pre-war pattern in several other important ways. That framework admittedly dictated the massive structural shift in production from agriculture to industry, and in population from countryside to town which, more than anything else, distinguishes the post-war period. Yet the large factories and agricultural complexes that now dominate the production process have a special significance, aside from their state ownership and socialist management. Their dimensions are also a striking departure from the small scale of production in the pre-war Bulgarian economy.

These post-war enterprises are large, even by contemporary European standards. The contrast to the small farms and unusually small industrial enterprises that predominated during the pre-war period is even sharper. As we have seen in Chapters 2–5, average farm and enterprise size continued to shrink throughout the Second World War and even in the first few post-war years. The shift since 1948 to huge units of production undoubtedly afforded the Bulgarian economy some valuable economies of scale. How much the abruptness and the extent of that shift have reduced those economies remains a serious question. Clearly, training a previously peasant labour force to work within a large, inevitably impersonal enterprise was a task as formidable as providing purely technical education. Both tasks continue to confront the management of these factories and farms, who in turn need better training and greater opportunity to advance.

The post-war redirection of Bulgaria's foreign trade toward the Soviet Union and the newly Communist states of Eastern Europe is a second major departure, socialism and CMEA aside, from the pre-war pattern. The pre-1914 and inter-war dependence on Central Europe in general, and on Germany in particular, has now shifted eastward. The difficulty of selling manufactured exports to a highly developed and united German economy was an enduring barrier to pre-war Bulgarian industrialisation. Access to the less demanding Soviet market for manufactures has undoubtedly aided post-war industry, as have Soviet supplies to Bulgaria of oil, iron ore, cotton and other raw materials of which pre-war Germany was itself an importer. The extensive export of Bulgarian canned foods, cigarettes and wine to the USSR, and also to Poland, Czechoslovakia and East Germany, may be traced in part to a climatic advantage over these north-eastern economies. But this was an advantage also enjoyed over a united Germany. The general weakness of socialist agricultural performance, with the exception of Hungary, seems as important as climate in accounting for Bulgarian food exports to the USSR and the rest of Eastern Europe.

The Soviet Union's political desire to strengthen its new camp of Communist states after the Second World War admittedly provides the best explanation for Bulgaria's better access than before 1944 to foreign investment, at least during the 1960s. The USSR took virtually nothing in post-war reparations, and protected the Bulgarian economy from the kind of Allied claims that hung over it throughout the 1920s (see Chapter 2). Soviet credits at low interest furnished over one-quarter of Bulgarian industrial investment capital during the 1950s, compared with the 10 per cent provided by the European Great Banks during the last pre-1914 decade and 15 per cent during the 1920s. Yet the renewed Bulgarian turn to Western European bank credits to cover machinery imports during the 1970s offers another reminder that the small Balkan economy's search for best technological practice, frustrated by pre-war capitalist neglect, faces a post-war socialist problem: that is, an Eastern European record of technical modernisation and managerial initiative that has consistently lagged behind Western best practice. To this deficiency, add Soviet economic difficulties since the mid-1970s, which have held back new investment and now cheap oil supplies from the USSR. Even overdue Soviet reforms may not fill these gaps.

Unbroken political continuity constitutes a third distinguishing feature of the post-war Bulgarian economy. The four decades of Communist rule since 1944 are by far the longest period of stable government in modern Bulgarian history. The shorter period (from 1886) of Prince Ferdinand's rule before the First World War must be further subdivided by the frequent changes of regime and the political divisions that he himself encouraged. Then came the turmoil of two world wars, framing two contentious inter-war decades. Each included a *coup d'état* (1923 and 1934) and other changes in government. The post-war period is not only longer, but has also been unmarked by the violent opposition that has convulsed each of the Communist regimes to the north (East Germany in 1953, Hungary and Poland in 1956, Czechoslovakia in 1968, and Poland again in 1980). Nor have the three decades of Todor Zhivkov's tenure as party First Secretary resulted in the sort of personality cult and defensive isolation around a long-time leader that has paralysed the prospects for economic reform in neighbouring Romania. Without Zhivkov's personal willingness to accept some measure of collective leadership and to face some shortcomings honestly, Bulgaria's nearly continuous pursuit of economic reform would not have been possible.

To the considerable extent that the most recent reforms rest on personal agricultural plots, small new manufacturing firms for consumer goods, and employee responsibility for enterprise profits, they depart from the Soviet pattern of central planning and ministerial supervision for large units of production. These reforms also appear to reverse a fourth pre-war tendency, the Bulgarian population's readiness to withdraw from the regular market-place in reaction to state pressure for increased sales. Recall the peasant cuts in grain marketed before and after the First World War, in favour of tobacco and then away from it again. During the Second World War, they favoured fruit and vegetables in order to escape requisitions by the grain-trading monopoly, Hranoiznos (see Chapters 1–4). The urban black markets of both world wars were part of the same pattern of behaviour. The recent official encouragement of small-scale production and individual initiative, albeit within a socialist framework, has helped to keep the Bulgarian black market one of the smallest in Eastern Europe.

Despite these four significant departures from the pre-war pattern, beyond the socialist framework and CMEA, the present Bulgarian economy still contains important elements of historical continuity. The Ottoman legacy of imperial subjugation does seem largely to have faded, despite its continuing reputation among some Western observers as an easy explanation for what is presumed to be Bulgaria's simple, subordinate relationship to the Soviet Union. Much more relevant to the complex economic relationship with the USSR and to the economy's internal dynamics are legacies from the first decades of Bulgaria's past century as a nation-state, governed from Sofia within roughly the present borders. These historical legacies apply respectively to the state's role in industry, agriculture and foreign trade, and to the limits of population growth and political leadership.

Modern industry has grown since the turn of the century at a faster rate and with greater state participation than in most other European economies. In part, mechanically powered production grew at rapid rates before and after the First World War, as well as since the Second World War, because it began from the smallest industrial base of any Eastern European country. The large artisan sector of the nineteenth century served only to impede the rise of modern industry. Chapter 1 has described artisan hardships from the loss of the large Ottoman market and from the movement of its labour force to lowland agriculture. The state's tax and tariff exemptions to encourage modern industry, dating from the 1890s, were therefore essential to attracting capital away from booming agricultural exports to Central Europe. Easier access to these state encouragements in the capital city started a concentration of industry in Sofia that continued throughout the inter-war period. Actual state ownership had begun by 1900 with the railway system and the major coal mines. By 1939, the state's share of industrial production had risen to 15 per cent, if co-operative output is included.

Also paving the way for post-war nationalisation was the state's role in the pre-1914 cartel for tobacco processing and its leverage in several industrial branches by the 1930s, which was achieved primarily through price controls and government purchase. Military mobilisation for the Second World War expanded this leverage. The fact that state controls and the earlier encouragement laws promoted the appearance of small-scale private firms and did not aid the larger ones only strengthened the post-war Communist argument for nationalisation. Private enterprise could not promise the badly needed economies of large-scale production. The present Bulgarian problem is how to reduce the size of overly large enterprises, whose diseconomies of scale are partly the result of the abrupt post-war transition from small to large.

Bulgaria's agricultural tradition during the twentieth century is perhaps the most progressive, by any definition, among all the Eastern European economies. The rapid growth of grain cultivation and export before the First World War was followed by a diversification first to tobacco, and then to other crops and to livestock. Aggregate growth continued at a significant rate during the difficult interwar years, in contrast to low rates for its Balkan neighbours. Under the primary leadership of Aleksandur Stamboliiski's Agrarian Party, a massive Bulgarian co-operative movement created its own credit network. This network linked up with the region's earliest (1903) and largest agricultural bank to provide peasant smallholders with a relatively large — though still insufficient — amount of credit and other modernising assistance. By the 1920s, the bank was Bulgaria's largest financial institution. By the 1930s, however, it was affording co-operatives more credit than individual peasant smallholders. The existence of this semi-public bank and co-operative network also furnished the post-war Communist government useful leverage in the rapid nationalisation of agricultural trade after the Second World War.

Still greater state leverage came from the food export and tobacco monopolies. These had been set up by the early 1930s and greatly increased their powers during the war. The disappearance since then of a separate agricultural bank has combined with the familiar Communist emphasis on heavy industry to set increasingly restrictive limits on modernising investment in agriculture and food processing.

Bulgaria's foreign trade was already based on bilateral inter-state agreements at artificial exchange rates before the post-war turn to such agreements with the USSR and the other CMEA members. Similar, if less favourable, arrangements were signed with Nazi Germany during the 1930s, as detailed in Chapter 3. They accounted for over one-half of exports and imports before the Second World War began. This early departure from free trade at fixed rates of convertible exchange was not really the result of close political alliance with Germany. It derived instead from the 1930s depression, and also from unpromising Bulgarian experiences with free Western European markets and with an overvalued currency before and after the First World War (see Chapters 1 and 2). High protective tariffs during the 1920s had failed to improve this bad bargain with free multilateral trade. The most heavily protected branches did not respond with rapid growth, while prohibitive rates discouraged the growth of large-scale industry by facilitating the easy entry of small firms.

Bulgaria's bilateral agreements within CMEA undoubtedly created more exports and imports, particularly during the general European trade boom of the 1960s, than did the less favourable pacts of the 1930s with Nazi Germany. But the general damage of such agreements to prices based on relative scarcity and to resource allocation based on opportunity cost has also been real. Most of Bulgaria's overdue price increases in 1979 were designed to repair some of this damage.

Higher export prices would hopefully reduce the deficit in Bulgaria's Western trade. Its short-run coverage had required heavy borrowing from Western banks. By 1976, as noted in Chapter 8, the 44 per cent ratio of debt service to hard-currency exports was the highest in Eastern Europe and greatly exceeded Bulgaria's inter-war peak. That ratio's sharp reduction since then has been the result of arbitrary cuts in Western imports and one-time Soviet oil deliveries for re-export. A long-term problem remains: how to trade for full comparative advantage with the free Western market, while still conducting a large majority of Bulgarian commerce under bilateral agreements at politically determined prices? Only area-wide economic reform around cost-based, market prices, including rates of multilateral currency exchange, can hope to resolve this contradiction.

The greatest historical pressure pushing current Bulgarian economic reform in the direction of cost-efficient use of all resources is the shortage of labour. Sofia, long the centre of Bulgarian industry, first experienced such a shortage before the First World War. Caused then by too little peasant immigration, the shortage forced up wages for some industrial workers and emboldened others, less fortunate, to join the young socialist movement. From the inter-war years throughout the 1950s, Sofia at least had a surplus of such labour. The roots of a longer-term labour shortage, which reached Bulgarian cities by the 1970s, lay instead in the countryside. The demographic losses of the First World War dovetailed with the poor agricultural prospects afterwards to push the rural Bulgarian birth rate downward. The decline has since become a national tendency. After a brief revival in the early 1950s, the Bulgarian birth rate turned down once again. The shock of collectivisation during the 1950s pushed many young peasants into the towns, where by the end of the decade too few factory jobs awaited them (see Chapter 6). This decline toward zero population growth created an inevitable labour shortage, which therefore appeared first on the collective farms. Economic reform, not surprisingly, was launched there first, in the early 1960s, as we saw in Chapter 9. The urban labour surplus of the late 1950s had meanwhile encouraged the expansion of industry, so that its needs exceeded the workforce available by roughly 1970. By the 1980s, the urban supply of male labour was facing an absolute decline for the rest of the decade. Turnover none the less remains too high and efficiency too low.

Declining rates of growth for national income since 1980 (down to 3 per cent for 1983) may continue to fall, or at least to stagnate under current Western levels, if productivity per worker and — perhaps more important — per unit of investment do not move upward. Less Soviet (and Western) lending, plus the Bulgarian government's commitment to continuing the recent rise in domestic consumption, bars a return to the excessive, often inefficient investment of the first Five-Year Plans. Thus, the scarcities of both capital and labour, which plagued the growth of the pre-war Bulgarian economy, have now returned.

One final pre-war legacy can assist Bulgaria's political leadership in facing the challenge of growth and scarcity. That is modern Bulgaria's tradition, dating even from late Ottoman times, of pursuing national development more successfully through internal economic progress, rather than external political manoeuvres. Historically, such manoeuvres have all ended in disaster. Popular sentiment has long since lost confidence in them. Alternatives to a Communist government in the Soviet camp are not considered. But the idea of hard work for the home economy still enjoys widespread respect. More political leaders of the pre-war period were trained primarily in an economic discipline and advocated primarily economic programmes than in any neighbouring state. This primacy explains in part why agrarian and socialist movements appeared in Bulgaria first and why they were further developed there before the Second World War than anywhere else in south-eastern Europe. It remains a significant advantage in the present Communist pursuit of economic reform.

SELECT BIBLIOGRAPHY

Economic Development and History

Allen, Mark. 'The Bulgarian Economy in the 1970s', in Joint Economic Committee of the US Congress (JEC), *East European Economies Post-Helsinki*, (Washington, DC: US Government Printing Office, 1977), pp. 647–99.

Berov, Liuben, *Bulgaria's Economic Development through the Ages* (Sofia, 1980).

Feiwel, George R. *Growth and Reforms in Centrally Planned Economies: The Lessons of the Bulgarian Experience* (New York: Praeger, 1977).

Jackson, Marvin R. 'Bulgaria's Economy in the 1970s: Adjusting Productivity to Structure', in Joint Economic Committee of the US Congress, *East European Economic Assessment*, pt 1 (Washington, DC: US Government Printing Office, 1981), pp. 571–618.

—— 'Recent Economic Performance and Policy in Bulgaria', in Joint Economic Committee of the US Congress, *East European Economies: Slow Growth in the 1980s*, pt 3 (Washington, DC: US Government Printing Office, 1985), pp. 1–36.

Khristov, Khristo (ed.). *Iz istoriia na stopanskiia i sotsialniia zhivot v bulgarskite zemi* [From the History of Economic and Social Life in the Bulgarian Lands] (Sofia, 1984).

Lampe, John R. and Marvin R. Jackson. *Balkan Economic History, 1550–1950: From Imperial Borderlands to Developing Nations* (Bloomington, Indiana: Indiana University Press, 1982).

Pasvolsky, Leo. *Bulgaria's Economic Position after the War* (Washington, DC: Brookings Institution, 1930).

Popov, Nikola, (ed.). *Ikonomika na Bulgaria* [The Economy of Bulgaria], vols. I–VI (Sofia, 1969–80).

Royal Institute of International Affairs. *The Balkan States, vol. I: Economic* (London: Oxford University Press, 1936).

Todorov, Nikolai, *et al. Stopanska istoriia na Bulgariia, 681–1981* [The Economic History of Bulgaria, 681–1981] (Sofia, 1981).

Totev, A. Iu. *Sravnitelno izuchavane na bulgarskoto i jugoslovenskoto narodno stopanstvo* [A Comparative Inquiry on the Bulgarian and Yugoslav Economies] (Sofia, 1940).

Zagoroff, S.D. *The Economy of Bulgaria* (Washington, DC: Council for Economic and Industry Research, 1955).

Statistics

Bass, Elizabeth. *Bulgarian GNP by Sectors of Origin, 1950–1974*, OP-44 (New York: L.W. International Financial Research, 1975).

Chakalov, Asen. *Natsionalen dokhod i razkhod na Bulgariia, 1924–1945* [National Income and Outlay of Bulgaria, 1924–1945] (Sofia, 1946).

Council of Mutual Economic Assistance. *Statistical Yearbook of the CMEA Member States, 1962–* (London: IPC Industrial Press).

Jackson, Marvin R. 'Agricultural Output in Southeastern Europe, 1910–1938', *ACES Bulletin*, vol. XXIV, no. 4 (1982), pp. 49–87.
—— 'National Product and Income in Southeastern Europe before the Second World War', *ACES Bulletin*, vol. XXIV, no. 3 (1982), pp. 73–103.
Jackson, Marvin R. and John R. Lampe, 'The Evidence of Industrial Growth in Southeastern Europe before the Second World War', *East European Quarterly*, vol. XVI, no. 4 (1983), pp. 385–415.
Kiranov, P. 'Natsionalen dokhod na Bulgariia, 1939, 1944, 1945' [The National Income of Bulgaria, 1939, 1944, 1945], *Izvanredno izdanie na spisanie na Glavna Direktsiia na Statistika* (Sofia, 1946).
Lazarchik, Gregor. *Bulgarian Agricultural Production, Output, Expenses, Gross and Net Product, and Productivity, at 1968 Prices, for 1939 and 1948–1970*, OP-39 (New York: Riverside Research Institute, 1973).
Mollev, I. and A. Iu. Totev. *Tseni na zemedelskite produkti u nas, 1881–1934* [Prices of our Agricultural Products, 1881–1934] (Sofia, 1935).
Overseas Trade Department, Great Britain. *Economic Conditions in Bulgaria*. Annual, 1921–39 (London: HM Printing Office).
Popoff, Kiril. *La Bulgarie économique, 1879–1911* (Paris, 1920).
Shapkarev, Petur. *Proizvodstveniiat ritum v narodnoto stopanstvo na NR Bulgariia* [The Productive Growth of the National Economy of the PR Bulgaria] (Sofia, 1983).
Staller, George. *Bulgaria: A New Industrial Production Index, 1963–1972*, OP-47 (New York: L.W. International Financial Research, 1975).
Statisticheski godishnik na Bulgarskoto Tsarstvo, 1909–42 [The Statistical Yearbook of the Bulgarian Kingdom, 1909–42] (Sofia, 1910–43).
Statisticheski godishnik na NR Bulgariia, 1956– [The Statistical Yearbook of the PR Bulgaria, 1956–] (Sofia, 1957–).
Statisticheski spravochnik na NR Bulgariia, 1958– [Statistical Reference Book of the PR Bulgaria, 1958–] (Sofia, 1958–).
United Nations Economic Commission for Europe. *Economic Survey of Europe*. Annual since 1948. (New York: UN Secretariat).
Vunshna turgoviia na NR Bulgariia, 1939–1975 [Foreign Trade of the PR Bulgaria, 1939–1975] (Sofia, 1976).

General History and Politics

Pre-1944

Bell, John W. *Peasants in Power: Alexander Stamboliski and the Bulgarian Agrarian National Union, 1899–1923* (Princeton, New Jersey: Princeton University Press, 1977).
Crampton, Richard J. *Bulgaria, 1878–1918: A History*. East European Monographs no. 138. (New York: Columbia University Press, 1983).
Dimitrov, Ilcho (ed.). *Kratka istoriia na Bulgariia* [A Short History of Bulgaria] (Sofia, 1983).
Hoppe, Hans-Joachim. *Bulgarien: Hitlers einwilliger Verbundeter* (Stuttgart, 1979).
Jelavich, Charles. *Tsarist Russia and Balkan Nationalism: Russian Influence in the Internal Affairs of Bulgaria and Serbia, 1879–1886* (Berkeley, California: University of California Press, 1958).
League of Nations. *Chronology of Political and Economic Events in the Danubian Basin, 1918–1936, Bulgaria* (Paris, 1938).

Logio, George C. *Bulgaria Past and Present* (Manchester: Sherratt and Hughes, 1936).
Miller, Marshall Lee. *Bulgaria during the Second World War* (Stanford, California: Stanford University Press, 1975).
Oren, Nissan. *Bulgarian Communism: The Road to Power, 1934–1944* (New York: Columbia University Press, 1971).
Rothschild, Joseph. *The Communist Party of Bulgaria: Origins and Development, 1883–1936* (New York: Columbia University Press, 1959).

Post-1944

Auty, Phyllis. 'Bulgaria', in R.R. Betts (ed.), *Central and South East Europe, 1945–1948* (Westport, Connecticut: Greenwood Press, 1971; reprint from 1951), pp. 25–51.
Boll, Michael M. *The Cold War in the Balkans: American Foreign Policy and the Emergence of Communist Bulgaria, 1943–1947* (Lexington, Kentucky: University of Kentucky Press, 1984).
Bokov, Georgi (ed.), *Modern Bulgaria* (Sofia Press, 1981).
Brown, J.F. *Bulgaria under Communist Rule* (New York: Praeger, 1970).
Isusov, Mito (ed.) *Problems of the Transition from Capitalism to Socialism* (Sofia, 1975).
Oren, Nissan. *Revolution Administered: Agrarianism and Communism in Bulgaria* (Baltimore, Maryland: Johns Hopkins Press, 1973).

Industry

Gerschenkron, Alexander. 'Some Aspects of Industrialization in Bulgaria, 1878–1939', in Gerschenkron, *Economic Backwardness in Historical Perspective* (New York: Praeger, 1965), pp. 233–98.
Kaser, Michael C. 'The Industrial Enterprise in Bulgaria', in Ian Jeffries (ed.), *The Industrial Enterprise in Eastern Europe* (New York: Praeger, 1981), pp. 84–94.
Marinov, Khristo. 'Geografsko razpredelenie na promishlenostta v Bulgariia mezhdu dvete svetovni voini' [The Geographic Distribution of Industry in Bulgaria between the Two Wars]. *Trudove na V.I.I. Karl Marx*, vol. 1 (1965), pp. 3–41.
Palairet, Michael. 'The Decline of the Old Balkan Woollen Industries, 1870–1914', *Vierteljahrschrift für Sozial und Wirtschaftsgeschichte*, vol. 70, no. 3 (1983), pp. 331–62.
Simov, B., B. Blagoev and O. Oslianin. *Vustanovanie i razvitie na promishlenostta v NR Bulgariia, 1944–1948* [The Growth and Development of Industry in the PR Bulgaria, 1944–1948] (Sofia, 1968).
Vasliev, V. 'Za glavnite faktori na industrialniia podem (1925–1929)' [Concerning the Main Factors of the Industrial Boom (1925–1929)], *Izvestiia na institut za istoriia*, vol. XI (1962), pp. 83–100.

Agriculture

Michael, Louis G. *Agricultural Survey of Europe: The Danubian Basin, Part 2:*

Rumania, Bulgaria and Yugoslavia. Technical Bulletin no. 126 (Washington, DC: US Department of Agriculture, 1929).
Minkov, M., *Poiava i razvitie na kooperativno zemedelie v Bulgariia pri usloviiata na kapializma* [The Appearance and Development of Co-operative Agriculture under Conditions of Capitalism] (Sofia, 1968).
Moloff, J.S. 'Bulgarian Agriculture', in O.S. Morgan (ed.), *Agricultural Systems of Middle Europe*, (New York: AMS Press, 1969; reprint from 1933), pp. 41–85.
Organization for Economic Co-operation and Development. *Prospects for Agricultural Production and Trade in Eastern Europe*, vol. 2, *Bulgaria* (Paris: OECD, 1982).
Sanders, Irwin T. *Balkan Village* (Lexington, Kentucky: University of Kentucky Press, 1949).
Wädekin, Karl-Eugen. *Agrarian Policies in Communist Europe* (London: Martinus Nijhoff, 1982).
Wiedemann, Paul. 'The Origins and Development of Agro-Industrial Complexes in Bulgaria', in Ronald A. Francisco, Betty A. Boyd and Roy D. Laird (eds.), *Agricultural Policies in the USSR and Eastern Europe* (Boulder, Colorado: Westview Press, 1980), pp. 98–109.
Zagoroff, S.D., J. Vegh and A.D. Bilimovich. *The Agrarian Economy of the Danubian Countries, 1933–1945* (Stanford, California: Stanford University Press, 1955).

Foreign Economic Relations

Bozhkov, Triatko. *Nauchno-teknicheska i proizvodna integratsiia mezhdu NR Bulgariia i SSSR* [Scientific-Technical and Productive Integration between the PR Bulgaria and the USSR] (Sofia, 1976).
Katsarkova, Vera. 'Ograbvaneto na Bulgariia ot germanskiia imperializm' [The Exploitation of Bulgaria by German Imperialism] *Trudove na V.I.I. Karl Marx*, vol. II (1969), pp. 164–223.
Khristov, Khristo, V. Hadzinikolov *et al. Bulgarskoto-Germanski otnosheniia i vruski* [Bulgarian-German Relations and Ties], vols. I–III (Sofia, 1972–81).
Lamb, Deborah A. 'Bulgaria: Performance and Prospects in Trade with the United States and the West', in Joint Economic Committee of the US Congress, *East-West Trade: The Prospects to 1985* (Washington, DC: US Government Printing Office, 1982), pp. 22–50.
Markov, Georgi. *Bulgaro-germanskite otnosheniia, 1931–1939* [Bulgarian-German Relations, 1931–1939] (Sofia, 1984).
Nikov, Angel. *Bulgaro-suvetski otnosheniia, 1944–1948* [Bulgarian-Soviet Relations, 1944–1948] (Sofia, 1978).
Nikova, G. 'Bulgaria and Cooperation of the CMEA Member-Countries in Production Planning, 1949–1956', *Bulgarian Historical Review*, vol. 12, no. 2 (1984), pp. 3–17.
Ofer, Gur. 'Growth Strategy, Specialization in Agriculture, and Trade: Bulgaria and Eastern Europe', in Paul Marer and John Michael Montias (eds.), *East European Integration and East-West Trade* (Bloomington, Indiana: Indiana University Press, 1981), pp. 299–313.
Popisakov, Grigor. *Ikonomicheska otnosheniia mezdhu NR Bulgariia i SSSR* [Economic Relations between the PR Bulgaria and the USSR] (Sofia, 1968).
Todorova, Tsvetana. *Diplomaticheska istoriia na vunshnite zaemi na Bulgariia, 1878–1912* [A Diplomatic History of Bulgarian Foreign Loans, 1878–1912] (Sofia, 1971).

Finance and Banking

Berov, Liuben. 'Le capital financier occidental et les pays balkaniques dans les années vingt', *Etudes balkaniques*, vols. 2–3 (1965), pp. 139–69.
Chakalov, Asen. *Formi, razmer i deinost na chuzhdiia kapital v Bulgariia, 1878–1944* [The Form, Extent and Activity of Foreign Capital in Bulgaria, 1878–1912] (Sofia, 1962).
Lampe, John R. 'Finance and pre-1914 Industrial Stirrings in Bulgaria and Serbia', *Southeastern Europe*, vol. II, no. 1 (1975), pp. 23–52.
Nedelchev, Kiril. *Parichnoto delo v Bulgariia, 1879–1940* [Monetary Affairs in Bulgaria, 1879–1940] (Sofia, 1940).
Totev, Totiu, 'Bankovata sistema na NR Bulgariia sled sotsialisticheskata revoliutsiia', [The Banking System of the PR Bulgaria after the Socialist Revolution], *Finansi i kredit*, vol. XXXI, no. 7 (1981), pp. 42–52.
Wynne, William H. *State Insolvency and Foreign Bondholders*, vol. II (New Haven, Connecticut: Yale University Press, 1951), pp. 530–74.

Population and Living Standards

Berov, Liuben. *Polozhenieto na rabotnicheska klasa v Bulgariia pri kapitalizm* [The Position of the Working Class in Bulgaria under Capitalism] (Sofia, 1968).
────── 'Realnite dokhodi i potreblenieto na trudeshite i seliana v Bulgariia prez 1938–1969' [Real Income and the Consumption of Workers and Peasants in Bulgaria, 1938–1969], *Problemi na truda*, vol. I (1970), pp. 23–34.
Isusov, Mito. *Rabotnicheska klasa v Bulgariia, 1944–1947* [The Working Class in Bulgaria, 1944–1947] (Sofia, 1971).
Jackson, Marvin R. 'Changes in the Ethnic Content of the Balkan National Populations, 1912–1970', *Faculty Working Papers*, Department of Economics, EC 83/84–20, Arizona State University. Forthcoming in *Southeastern Europe*.
Lampe, John R. 'Interwar Sofia vs. the Nazi-style Garden City: The Struggle over the Muesmann Plan', *Journal of Urban History*, vol. II, no. 1 (1984), pp. 39–62.
Radulov, L. *Naselenie i ikonomika* [Population and Economics] (Sofia, 1981).
────── (ed.), *Sotsialno-ikonomicheska politika na Bulgarskata durzhava, 681–1981* [The Socio-Economic Policy of the Bulgarian State, 681–1981] (Varna, 1981).

Economic Reform

Feiwel, George W. 'Economic Reform in Bulgaria', *Osteuropa Wirtschaft*, vol. 24, no. 2 (1979), pp. 71–91.
────── 'Economic Development and Planning in Bulgaria in the 1970s', in Alec Nove, Hans Hermann Höhemann, Gertrud Seidenstecher (eds.), *The East European Economies in the 1970s* (London: Butterworths, 1982), pp. 222–37.
Nikov, Ivan. *Ikonomicheskiiat podkhod* [The Economic Approach] (Sofia, 1983).
Vogel, Heinrich. 'Bulgaria', in Hans-Hermann Höhemann, Michael C. Kaser and Karl C. Thalheim (eds.), *The New Economic Systems of Eastern Europe* (Berkeley, California: University of California Press, 1975).

Select Bibliography

Periodicals

Bulgarian

Bulgarian Historical Review
Economic Outlook of the PR Bulgaria (Bulgarian Chamber of Commerce and Industry)
Finansi i kredit
Ikonomicheska misul
Ikonomicheski zhivot
Ikonomika na selsko stopanstvo
Izvestiia na Bulgarska Narodna Banka (1924–40)
Izvestiia na Ikonomicheski Institut na BAN
Novo vreme
Planovo stopanstvo
Problemi na truda
Rabotnichesko delo
Spisanie na bulgarskoto ikonomichesko druzhestvo (1896–1940)

Western

East European Economies (New York: IASP Press)
The Economist Intelligence Unit, *Bulgaria and Romania* (London)
Radio Free Europe Research, *Bulgarian Situation Report* (Munich)
Wharton Econometric Forecasting Associates, *Centrally Planned Economies, Current Analysis* (Washington, DC)

INDEX

Agrarian Union 30
agriculture 16, 23-4, 27-9, 43, 60, 228
  consolidation of strips 57, 83, 112
  control of 43, 107-10, 127
  credit 83-4
  delivery prices 43, 148
  exports 31, 56, 84, 90, 108, 114, 177, 181, 227
  see also grain; tobacco
  fertiliser 58, 110, 170-2, 190, 206
  irrigation 84, 170, 172, 174
  marketing 8-9, 27-8, 56, 59, 80, 123, 147-8
  Ministry of Agriculture 57, 60, 146, 208
  personal plots 28, 210-12, 226
  prices 78, 85
  surplus, forced collection of 71
  trade, nationalisation of 124-7
  use of iron ploughs 28, 59-60
  use of tractors 170
  workers for Soviet Union 150
  see also banks; Agricultural Bank; collectivisation; crops; growth; industrial crops; machine-tractor stations
Agro-Industrial Complexes (APK) 206-9
Agro-Industrial Union 208
agronomists 43
  of Sofia University Faculty of Agronomy 58
Alexander III, Tsar 19
Alexander of Battenberg, Prince 19
Allied Control Commission 122-3, 125
Alton, Thad 161
Andropov, Yuri 219
army purge 121
artisan manufacture 20, 22-3, 39, 41, 71, 94, 132, 150, 227
Austria 34, 61, 88-9, 189
autarchy, East European 3
authoritarianism 79, 106
automobiles, private 192-4

Bagrianov, Ivan 107

Bairoch, Paul 1, 14, 24
balance of payments 73, 196
  deficit 114, 189-90, 205
Balkancar enterprise 168, 184, 190
Balkan customs union 88, 120-30
Balkan Tobacco Office 88
Balkan War, First 20
Balkan War, Second 13, 17, 20, 42
bank assets 31, 33-4, 93
banks, Bulgarian 37, 130
  Agricultural Bank (Bulgarska Zemedelska Banka) 30-1, 34-5, 58, 60, 67, 81-3, 92-3, 99, 114, 126, 128, 131, 205, 228
  Bulgarska Narodna 31-4, 91-2, 130-1
  Bulgarski Kredit 92, 130
  Central Co-operative Bank 30, 58, 81, 83, 92
  co-operative 30, 58
  Foreign Trade Bank 205
  Industrial Bank 205
  Investment Bank 205
  National Bank 31, 33, 37, 63-4, 70, 204-5, 215-16, 218
  reserve ratio 34, 63, 115
  see also currency; note issue
  popular banks 36, 51, 92, 130-1
  Turgovska Banka (Ruse) 34
banks, foreign 34, 93, 130
  Balkanska and Generalna 66
  Berliner Disconto-Gesellschaft 44, 64
  Deutsche Bank 32
  Franco-Belge 130
  Franco-Bulgarska 68
  Generalna 34
  Kreditna Banka 66, 68, 93
  Paris Bas 89, 130
  West European 66-8, 70, 93, 225
barter trade 88-9, 127
birth rates 55, 74, 86, 159, 229
black market 114, 126, 226
Black Sea 17, 29, 52, 186, 189, 203
Blagoev, Dimitur 35, 40
bombing raids, Anglo-American 106-7, 116, 133

237

Index

Boris, Tsar 79–80, 105–6, 107
Bozhilov, Prime Minister 107
Brittan, S. 6
Bucharest 28
Budapest 17, 28
Budget, State
  agriculture 60, 66
  debt burden, *see* foreign loans
  defence 34, 100
  deficits 29, 44, 62, 64–6
  education 66
  expenditure 115, 153, 205, 220n15
  revenues 33, 42, 61, 65, 101, 205
  state bureaucracy and 65, 117
Bulgarian Agrarian National Union
    (BZNS) 29–30, 39–40, 49–50,
    57–9, 123
  Pladne Agrarians 78–9
  Vrabcha Agrarians 78
Bulgarian Communist Party (BKP),
    (1918–27) 49–50, 106–7,
    121–3; (1948–) 140–1, 149,
    158, 201, 207, 217, 223, 226
Bulgarian Economic Society (BID)
    39, 56
Bulgarian Industrial Association 217
Bulgarian Workers' Party (1927–48)
    50, 79
Bulgarplod enterprise 209
Burgas 29, 32

Canada 29
capital 14, 230
  investment (accumulation) 29,
    35–7, 40, 71, 156, 161, 223
  agricultural 169, 207–8
  funds, competitions on 218
  Western 67–8, 190
  output ratio 209
  productivity 10, 86–7, 158, 163–5,
    174
  rate of accumulation 143, 164–5
  state investment 144, 216
cartels 39, 69, 99
  sugar 43–4, 69–70, 82, 99
  tobacco 69, 227
cement production 43–4, 72, 100,
    190, 195
  Granitoid plant 68, 73, 134
Central Co-operative Bank 30, 58,
    81, 83, 92
Central Co-operative Union 210
Central Europe 54, 67
central planning, Soviet-style 139,
    141, 179, 185, 199, 200, 223, 226
  reformed 14–15, 156
  transition to 17, 107, 121, 136,
    139, 141, 152
Chakalov, Asen 71, 94
chemicals, production of 72, 117,
    153, 168, 174, 190
Chervenkov, Vulko 123, 140–1, 144,
    149, 151, 153
chiflik 21–3
China 149, 151
chorbadzhiia 21
Churchill, Winston 13
Claudius, Karl 113
clearing agreements 91
  bilateral 87–8
  with Germany 87–8, 92
coal 17, 41, 68, 115, 117, 145
  imports 41, 72, 168, 223
  lignite 17, 192
  Pernik mines 41, 44, 100, 134, 144
Coca-Cola 190, 203
collective farms (TZKS or *kolkhoz*)
    125, 146, 152–3, 191, 200, 203,
    206–7, 210, 230
  Model Charter for 147
collectivisation 15, 71, 124–6,
    140–1, 143, 146–9, 169, 206,
    228–9
Cominform 135
Compulsory Labour Service
    (*trudovaki*) 58–9
consumer goods 97, 134, 188, 190,
    217, 223, 226
  durables 192, 194
consumption 191, 194, 196, 230
  meat 192, 196
  personal 97, 191–2
co-operative banks 67, 130
co-operatives 51, 82, 126–7, 228
  credit 30, 58, 81, 83–4
  marketing 82–3
  producers 30–1, 40, 68, 115
copper mining 17, 44, 68, 115
Corecom shops 193
cotton 82, 84, 99, 109
  imports 22, 94, 113, 117, 128, 134,
    173, 186
  production 60, 74, 84, 94, 110,
    150, 173
Council for Mutual Economic
    Assistance (CMEA) 150–1, 177,
    183–7, 224, 229
  Specialisation Agreements 183, 185

Index 239

Council of Ministers 158, 212
counterplanning 214
credit 29–31, 92, 205
  long-term 37, 83, 89, 205
  mortgage 30, 58–9
  short-term 22, 30–1, 36–7, 58,
    66–7, 70, 83, 205
crops
  diversification of 85–6, 87, 112,
    228
  land distribution by 26
  livestock ratio to 171–4
  production of 27–9, 53, 84–6,
    126–7, 143–4, 151, 153,
    171–4, 207, 226
  wartime 110–12, 169–70
  see also industrial crops
Cuba 185
currency
  convertible 62–3
  depreciation 45, 60, 62–3, 72, 89,
    204
  exchange rate 35, 44, 61–3, 91,
    228
  premiums 92
  money supply 62
  note issue 33, 60–1, 63, 114–15,
    131
Czechoslovakia 49, 88–9, 129, 151,
  201–2, 212, 225–6

dairy production 84
Danube River 17, 29
death rates 55, 159
debt, relief from 83
  see also under foreign loans
decentralisation 10, 199, 212
demand-management policies 6
Democratic Concord 50, 78
Democratic Party 50, 78
Denmark 28, 71
Depression (1929–33) 3, 8, 74, 78,
  80, 93, 98, 228
development 8, 10, 15, 68, 70, 142,
  156, 230
Dimitrov, Georgi 122–3, 125, 130,
  135, 140
Dimitrovgrad 142
discount rate 64
Dobrudja 18n5, 20, 42, 51–2, 104,
  108, 112

Eastern Europe
  growth 5
  reform 199, 201

education 19, 22, 58
  higher 195
  literacy 28
  primary 19, 193
  secondary 58, 66
  technical 195
eggs 27, 31, 54, 67, 90, 108–9, 172,
  186, 206
elections 30, 49–50, 78–80, 122
electric power 58–9, 128, 142–3,
  145–6, 153, 180, 184, 223
electronics 168, 174, 180, 185–6,
  190, 212–13
emigration 17
employment
  industrial 37–8, 68, 70, 97,
    132–3, 153
  state 37, 64, 100, 121, 134–5
environment 10
Europe 1–8, 14, 32, 66–8
European Economic Community 183
exchange rate, see currency
exports 14, 78, 92, 108, 128, 151,
  168, 177, 186, 224
  diversification 27
  growth 74, 94, 178
  prices of 108, 114, 145, 178, 181,
    229
  real 25
  state controls 81–2
  surplus 23, 25, 44, 113
  see also agriculture; exports

fascism 80, 106
Fatherland Front (OF) 106–7,
  121–2, 124
Ferdinand of Coburg, Tsar 19–20,
  30, 32, 39, 49, 112–14, 123, 226
fertility 16
Filipov, Grisha 202
Filov, Bogdan 106, 109
First World War 2, 8, 20, 42–5
Five-Year Plans 9, 139–54
  1942–6 109
  1949–52 141–3, 147–8
  1953–7 144, 148–9, 185
  1958–60 141, 146, 149–54, 200
  1961–5 139, 164
  1966–70 164
flax 108–9
flour sales surcharge 81
food
  consumption 191–2, 212
  exports 45, 71, 90–1, 94, 145, 168,

181, 186, 188, 191, 224–5, 228
processing 117, 151, 153, 206–7, 223–4
growth of 69, 95, 150
shortage 125–6, 134–5, 207
forage crops 172–4, 210–12
foreign loans 23, 32, 34, 63–4, 67
debt burden 32–3, 61, 64, 131, 189
debt service 34, 42, 64, 88–9
ratio to export earnings 64, 189, 229
France 33, 131
League of Nations 62
see also under Germany; Union of Soviet Socialist Republics
foreign trade 88–9, 92, 129–30, 150–2, 156, 177, 228–9
balance of 179–83
direction 54, 129, 152, 188, 225
Directorate for 109–10
growth of 20, 178–9, 184
structure 178, 180
terms of trade 115, 181–2
turnover 14–15, 178, 185
see also under Germany; Great Britain; Union of Soviet Socialist Republics; Western Europe
France 32, 50, 52, 62, 79, 89, 190
bank affiliates 34, 66–7
gold franc 62
loans from 33, 61
Friedman, Philip 91
fruit
exports 90–1, 108, 128
growing 110–12, 126, 173–4, 207, 209, 211
pulp processing 91, 108, 117, 126
full employment in Europe 4–5

Georgiev, Kimon 79
German-Bulgarian Industrial Commission 117
Germany 13, 80, 88
bank affiliates 34, 66
Democratic Republic (GDR) 199–201, 226
loans from 32, 44, 64, 89
trade relations with 9, 44–5, 54, 87–91, 99–100, 113–14, 229
East (GDR) 151, 184, 213, 225
West (FRG) 5, 151, 187–8

wartime deliveries to 107–9, 114
Central Agency for Special Requisitions (ASR) 109
Directorate for Civilian Mobilisation (DGM) 109–10, 116, 126
Directorate for Special Requisitions 43, 45, 109
Gerschenkron, Alexander 94
Geshov, Ivan Estatiev 40–1
Gichev, Dimitur 79
gold exchange standard 62–3, 78, 80
gold standard 1, 3, 32, 63
Gorubso joint enterprise with the USSR 150
grain 16, 67, 111, 113
consortium 57, 59
consumption 27, 52–3, 173
corn 25–7, 29, 172
exports 23–9, 52–4, 81, 87, 108–9, 228
composition of 53
imports 43
prices 25–6, 28–9, 37, 57, 82
production 24, 26–7, 31, 51–3, 85, 110–11, 172, 211, 228
rye 109, 172
storage 52, 57
wheat 15, 23, 25–6, 28–9, 31, 51–2, 109, 172–3
yields 29, 52, 84, 172–3, 207, 212
Great Britain 50, 67
trade relations with 40–1, 52, 89, 91
Greece 62, 88–9, 94, 105, 110, 113, 131, 180, 189, 207
bankruptcy of 1897 33
Gross Domestic Product (GDP) 162
Gross National Product (GNP) 14, 24, 161–2
Alton estimates 161
Gross Social Product 25
growth 8, 10–13, 17, 24, 81, 158, 199, 201, 218
agricultural 20, 27, 84–5, 144, 150–1, 162–3, 169–71, 212, 228
cycles 5–6
export-led 14
extensive 9
industrial 9, 35, 68–72, 93–8, 115–16, 227, 230
under communism 144–5, 150, 153, 156, 162–4

Index 241

intensive 9–10, 156, 163
rate, real 25
see also Eastern Europe
Gypsies 17

hemp 82, 109
Higher Economic Council (VSS) 126
horsepower 36, 68, 70, 116
housing 10, 59, 65, 151, 191, 193–5
Hranoiznos (grain export agency) 81–2, 84, 109, 111, 126–7, 226
Hungary 81, 91, 117, 158, 164, 173, 177, 217, 225–6
economic reform 199, 210

immigration 49, 160
import(s) 25, 38, 100, 117
elasticity 178
growth 179
of inputs 41, 73, 94, 99, 117, 145, 179, 186
licences 88, 91, 190
prices 114, 178, 181
substitution 9, 38, 72, 74, 93–4, 96, 99, 167–8, 179, 190, 193
surplus 89, 182
incentives 10, 145, 153–4, 205, 215, 218
Incentives Fund, Additional Material 216
income per capita, European 1, 7–8
independence 13, 19–20
Industrial-Agricultural Complex (PAK) 207
industrial crops 52–3, 57, 59, 84–5, 110, 171, 173–4
industrial firms, number of sizes of 95–6, 116–17, 132, 146, 153
industrialisation 93, 118, 140, 225
debate 38–45
industry 15, 69
combines 206, 212–14
DSO (state economic organisations) 204, 212, 214
employment 37–8, 68, 70
encouragement laws 41, 72–3, 227
heavy (producers' goods) 40, 69, 139, 141–6, 150–1, 223, 228
inputs
imports of 41, 73, 94, 99, 117, 145, 179, 186
prices 99, 165
labour force 139, 143, 145–6, 151, 153, 156, 163, 185, 224

light (consumer goods) 10, 143, 223
management 10
nationalisation of 9, 132–6, 227
output 10, 36, 68, 94–5, 116, 143–4
productivity 156, 165, 230
profitability 69, 201–2, 215–18, 226
'saturated' 98–9
state control 98–100
state-encouraged 9, 20, 37, 59, 68, 72–3, 95, 98–9, 227
structure of 166–9
subsidiaries 213
Union of Bulgarian Industrialists 133
wartime 43–4, 115–18
see also under growth
inflation 65, 112–15
European 4, 6–7
infrastructure 10, 28, 34
insurance societies, agricultural 59
interest rates 32–4, 37, 216
Internal Macedonian Revolutionary Organisation (IMRO) 13, 50, 65, 78–9
International Labour Organization 192
iron ore, *see* metallurgy
iron ploughs 28, 59–60
ishleme 22, 134, 186
Istanbul (Constantinople till 1908) 20, 21–2
Italy 89
Iugov, Anton 141

Jackson, Marvin R. 86, 94
Japan 189
Jews 107
joint-stock companies 37, 44, 68–9, 71, 94–5
joint ventures 150–1, 184
Law of 1980 190
with Japan 190

Kaser, Michael 81
Khrushchev, N. S. 140–1
Kioseivanov, G. 80, 106
Kostov, Traicho 123, 125, 132, 140, 201
Kulaks 146–8
Kunin, Petko 201

labour 22, 29
  abroad 185
  force, distribution of 160
  per enterprise 95
  productivity 161, 163, 167, 169, 201, 203, 208-9
  property 56-7
  shortage 29, 37, 160-1, 174, 195
    causes of 23, 118, 153, 185, 229-30
  trudovaki (Compulsory Labour Service) 58-9
  turnover 143-4, 161, 230
  week 58
  see also industry, labour force
land ownership 28, 57, 84
land reform 21, 23
  1920s 56-7, 59
  1946 125
League of Nations
  Interallied Debt Commission 62
  refugee loans 56, 63-4
Lenin, V. I. 61
Liapchev, Andrei 40, 50-1, 59-60, 63, 72-3, 78, 98, 100
Liberman, Evgeni 201
Libya 185
List, Friedrich 40
literacy 28
livestock 54, 144, 169-71, 174, 206, 212
  cattle 27, 108, 112, 127, 172
  hogs 27, 171
  poultry 27, 67, 171-2, 209
  sheep 27, 108
loans
  municipal 34
  state 63-4, 66, 131
  see also foreign loans
Lukov, General 80, 107

Macedonia 13, 19, 20, 44-5, 105-6, 112-13, 123
machinery 28-9, 31, 87, 108, 112, 170, 185, 187-8
  imports 59, 94, 179-80, 188
    from Germany 109, 114, 118, 184
    from USSR 125, 128, 143, 184
  production 35-6, 69, 117, 134, 145-6, 153, 166, 168, 174
machine-tractor stations (MTS) 128, 148
Malinov, Aleksandur 78, 82, 98

management 10, 217
  new system of 200-4, 213, 216
manufacturing 35-7, 68-9, 73-4, 115, 145, 214
  share in total output 8-9, 94-5
Mateev, Evgeni 201
mechanisation 23, 28, 87, 143, 227
  factory 20, 68, 84-5
metallurgy 68, 145, 166-7, 191, 223
  Burgas Complex 168, 224
  iron ore imports 167-8, 186, 223
  Kremikovtsi Complex 149, 167
  Lenin Complex 167, 184
  production 69, 72, 146, 174, 183
Midhat Pasha 23
Mihailovski, Angel 201
Mikhov, General 107
Military League 79
Ministry of National Economy 80
Ministry of State Planning 203
mobilisation, economic 109-10
  Directorate for Civilian Mobilisation (DGM) 43-4, 109-10, 116, 126
modernisation 79-80, 83-4, 153, 168, 228
monetary control in Europe 6
monetary policy 61, 131, 204
monetisation 20, 22, 29, 32, 35, 179
money supply 62
monopoly 82, 98, 214
monopsony 81, 109
most-favoured nation treatment (MFN) 88
municipal loans 34
Muraviev, Konstantin 107
Mushanov, Nikolai 78-9, 82-3, 98-9

National Bloc (Alliance Concord) 50, 78
national identity 19
national income 51-2, 71, 94, 115, 230
  net material product (NMP) 161-4, 178
  official estimates 162
New Economic Mechanism (NIM) 215-19
Norway 14
nuclear power 180, 184, 223

oil imports 114, 128, 179, 186-7, 224-5, 229

Orthodox Church 19
Ottoman Empire 19, 21–4, 40, 134, 227
Ottoman Public Debt 32

Palairet, Michael 38
parakende 28
partnerships 70–1
People's Courts 134
People's Guard and Militia 122
Pepsi-Cola 190
Pharmakhim enterprise 168
Pirin region 65, 70
planned socialist economies 5, 7
planning 5, 9–10, 100, 116
  see also central planning
Plovdiv 22, 150
plum brandy 82
Poland 21, 89, 225–6
Pomaks 23
population 13, 16, 113, 159
  birth rates 55, 74, 86, 159, 229
  density 16, 27, 56, 159
  emigration 17
  growth 23, 25, 27, 37, 55, 70, 86, 159, 229
  immigration 49, 160
  mortality rates 55, 159
  see also refugees
ports 23, 29, 34
poultry 27, 67, 171–2, 209
  see also eggs
power blocs 4
price(s) 37, 73–4, 115, 191, 193, 202, 205, 212
  control of 43, 99, 216, 227
  mechanism 218
  scissors 78
  see also export prices; import prices; inflation
Prices and Price Information, State Agency for 218–19
protectionist tariffs 3, 15, 40, 72, 228
pulp processing of fruit, *see under* fruit

railways 44
  construction 29, 32, 34, 39, 41, 60, 100, 135
  Deutsche Bank interest 32
  freight rates 73
  freight usage 143, 153
  Oriental Railway 32, 39, 41, 44

revenue 42
Rockefeller interest 32
rolling stock 52, 100, 143
trackage 52, 100
rainfall 16, 26
rationing 43
  bread 108–9
  meat 112
recovery, European 4–5
Reemstma (German trade organisation) 89, 108
reforms, economic 199, 200, 219, 224, 230
  agricultural 206–12
  financial 204–6, 215–19
  industrial concentration 212–14
  management 200–4, 213, 216
  market-oriented 199, 201, 229
refugee loans 56, 63–4
refugees 49, 56, 70
Renault 190
reparations 60–3, 89, 124, 131, 184, 225
  Interallied Reparations Commission 62, 80–1
resettlement loan 56
retail trade 115, 192, 202
Rhodopa enterprise 209
road construction 23, 65, 100
Romania 21, 29, 42, 49, 87, 104, 129, 184, 226
  comparisons with 24, 36, 40, 53, 57, 65, 73, 83, 91, 94
  under Communist rule 131, 161, 164, 171, 173, 192–3
rose oil 82
Rostow, W. W. 7
Ruse 29, 184, 218
Russo-Ottoman Wars (1828–9) 21; (1877–8) 19

savings banks 30
scarce resources 14, 16
Schacht, Hjalmar 89
Schweppes 190
Scientific-Productive Complex (NPK) 209
Second World War 4, 9, 105–18
Serbia 20, 24, 36–7, 40
sharecropping 27, 29
ship construction 134, 151
silk cocoons 27, 30–1, 82
small-scale enterprises 115–16, 132–3, 135, 145–6, 227

smallholders 123, 147−8
smallholdings 24, 27−30, 52, 56, 112
  percentage of arable land 57, 86
Social Democratic Party (1918−48)
  50, 122, 125
Socialist Party 49−50
social services 193−4
Sofia 9, 17, 41, 101, 106, 184, 218
  Mortgage Bank 93
  industrial concentration 41, 96−9
  population 38, 70, 195
  University 51, 87, 153, 195
soya beans 108, 172
stagflation 6
Stalin, J. V. 140−1, 144, 147
Stamboliiski, Aleksandur 13, 30, 39,
  50−1, 56−8, 79, 123, 151, 228
  regime of 56−8, 62, 64−5, 71−2,
  82
Stambulov, Stefan 20
standard of living 27, 177, 190−6
State control 9, 14, 15, 17, 19−20,
  41−5
State enterprise 100−1, 116−17,
  133−6, 227
State initiative 81
State Planning Commission (DPK)
  141, 215
Statistics, Main Directorate for
  (GDS) 142
structural change 8−10, 68, 166−9,
  171−4, 224
structural imbalance in Europe 3
Subranie (National Assembly) 19,
  39−40, 43, 80−1
Sugar Association (Bulgarski Zakhar)
  209
sugar beet 53, 72, 98, 173
  cartel 43−4, 69−70, 82, 99−100
  refining 31, 41, 68, 93
sunflowers 53, 84, 173
Sweden 14
Switzerland 129

TABSO joint enterprise with the
  USSR 151
tariffs 3, 15, 227−9
  customs revenue 34, 62, 64−5
  export 58, 62, 72
  import 40, 72, 99, 114
taxes 21, 23, 29−30, 131
  direct 65, 72, 148, 211, 227
  indirect 29−30, 34, 65, 205, 216
textiles 41, 68, 134, 184

exports 189
  production 38, 69, 98, 117, 145−6
Third World 24, 159, 182, 187
Thrace 105, 112−13, 179
Tito, Josip Broz 129
  Tito-Stalin split 130, 140
tobacco 15−16, 113, 126, 133, 228
  cartel 69, 227
  cigarette production 108, 186
  co-operatives 59
  exports 44, 52−5, 90, 108, 128,
  168
  prices 59
  production 52−5, 74, 84, 110, 150,
  173
  state monopoly 33−4, 42, 82
  United Tobacco Factory 130
tomatoes
  exports 90
  production 174
Totev, A. Iu. 86
tourism 189−90, 192
tractors, use of 170
  see also machine-tractor stations
trade, internal 22, 39
  retail 192, 202
  see also foreign trade
Treaty of
  Berlin (1878) 19, 24
  Neuilly (1919) 61, 72
  San Stefano (1878) 19
Triffin, R. 1−2
Tsankov, Aleksandur 50−1, 59−60,
  62, 65, 80, 106
Turkey 113, 117, 189
Turks 17, 18n6, 21, 24

unemployment 6−7, 133, 149−50,
  230
Union of Soviet Socialist Republics
  14, 61, 71, 94, 200
  joint companies 150−1, 184
  loans from 184, 189, 225
  oil imports from 128, 179, 186−7,
  189
  political relations with 10, 13, 105,
  184, 227
  Red Army 106−7, 124
  role on Allied Control Commission
  122
  trade relations 15, 79, 80, 84, 87,
  117, 127−30, 151, 167,
  183−7, 189, 195, 224−5
United Nations Economic

Commission for Europe
(UNECE) 161, 178
United States of America 3, 4
most-favoured nation treatment
189
role on Allied Control Commission
122-3, 125
uranium mining 151
urbanisation 37, 70, 159, 229-30

Varna, 23, 52, 134, 186
vegetable-growing 16, 26-7, 31, 57,
112, 126, 209, 211-12
exports 54, 108
growth of 84-5, 110, 173-4
Velchev, Damian 79
vineyards 85, 110, 173-4, 209
wine exports 128, 209

wages 23, 37, 134, 146, 171, 191,
215, 218
money 191
real 191, 196

wartime mobilisation
Committee for Public Welfare
(KZOP) 43, 51
Directorate for Civilian
Mobilisation (DGM) 43-4,
109-10, 116, 126
Western Europe, trade with 182, 184,
186-90, 229
Wool Directorate for Foreign Trade
38
working week 210-11
World Bank 161, 163

Yugoslavia 16, 18n5, 49, 62, 79, 87,
105, 110, 113, 131, 189
comparisons with 7, 53, 66, 83, 91,
94
under Communist rule 173, 207,
218
market socialism in 199

Zhivkov, Todor 141, 151-2, 158,
202, 213, 215, 226
Zveno 79, 83, 87, 92, 122